Disclaimer

The publisher of this book is by no way associated with the National Institute of Standards and Technology (NIST). The NIST did not publish this book. It was published by 50 page publications under the public domain license.

50 Page Publications.

Book Title: Technical Guidance for the Specification and Development of Homeland Security Simulation Applications

Book Author: Charles R. McLean; Sanjay Jain; Yung-Tsun T. Lee; Charles W. Hutchings;

Book Abstract: This document provides general guidance that is designed to help the U.S. Department of Homeland Security (DHS) Program Managers who may have limited expertise in simulation to better understand this technology. It is also intended to help the relevant professional community in developing better technical specifications for homeland security modeling and simulation (M&S) applications. Contractors outside of DHS typically develop homeland security M&S applications. Often the M&S applications that are developed today are difficult to extend for new purposes, integrate with other software systems, or used by persons other than the original developer/development team. Slight changes to a model or scenario may require major changes to a system and significant cost to the sponsor. Examples of topics (within the context of simulation) that are addressed in this report include: , Analysis of user needs and system requirements , Specification development: conceptual models, detailed designs, and interfaces , Modeling and simulation technologies appropriate to different problem domains , Use of proprietary commercial versus open source software , Project team and developer qualifications , Simulation engine , User interfaces , Inputs and output data types , Databases, data files, and translators , Documentation and training , Software risk management , Verification, validation, accreditation, and testing , Information security mechanisms , Distributed architectures and communications , Standards The technical guidance contained in this report is based on published literature, information available on the Internet, technical expertise, and personal experience.

Citation: NIST TN - 1742

Keywords: simulation applications; modeling; Homeland Security; conceptual models

NIST Technical Note 1742

Technical Guidance for the Specification and Development of Homeland Security Simulation Applications

Charles McLean
Dr. Sanjay Jain
Y. Tina Lee
Dr. Charles Hutchings

http://dx.doi.org/10.6028/NIST.TN.1742

National Institute of
Standards and Technology
U.S. Department of Commerce

NIST Technical Note 1742

Technical Guidance for the Specification and Development of Homeland Security Simulation Applications

Charles McLean
Dr. Sanjay Jain
Y. Tina Lee
Systems Integration Division
Engineering Laboratory
National Institute of Standards and Technology
Gaithersburg, MD 20899

Dr. Charles Hutchings
U.S. Department of Homeland Security
Washington, DC 20528

http://dx.doi.org/10.6028/NIST.TN.1742

May 2012

U.S. Department of Commerce
John E. Bryson, Secretary

National Institute of Standards and Technology
Patrick D. Gallagher, Under Secretary of Commerce for Standards and Technology and Director

National Institute of Standards and Technology Technical Note 1742
Natl. Inst. Stand. Technol. Tech. Note 1742, 107N pages (Month and Year)
http://dx.doi.org/10.6028/NIST.TN.1742
CODEN: NTNOEF

Technical Guidance for the Specification and Development of Homeland Security Simulation Applications

Charles McLean
National Institute of Standards and Technology
Gaithersburg, MD 20899

Dr. Sanjay Jain
George Washington University
Washington, DC 20052

Y. Tina Lee
National Institute of Standards and Technology
Gaithersburg, MD 20899

Dr. Charles Hutchings
U.S. Department of Homeland Security
Washington, DC 20528

May 2012

U.S. Department of Homeland Security
Washington, DC
and
National Institute of Standards and Technology
Gaithersburg, MD

This page left blank intentionally.

Acknowledgments

The U.S. Department of Homeland Security (DHS) Science and Technology Directorate sponsored the production of this material under Interagency Agreement HSHQDC-09-X-00357 with the National Institute of Standards and Technology (NIST). The work described was funded by the United States Government and is not subject to copyright.

Disclaimers

The findings expressed or implied in this report do not necessarily reflect the official view or policy of the U.S. Department of Homeland Security, U.S. Department of Commerce, or the United States Government.

Some software products may have been identified in context in this report. This does not imply a recommendation or endorsement of the software products by the authors or NIST, nor does it imply that such software products are necessarily the best available for the purpose.

This page left blank intentionally

Executive Summary

This document provides general guidance designed to help the U.S. Department of Homeland Security (DHS) Program Managers and other executives inside and outside of DHS. It is intended to provide useful information that may help personnel better understand the technologies used to maintain and support homeland security modeling and simulation (M&S) applications. It is also intended to help the relevant professional community in developing better technical specifications for homeland security modeling and simulation (M&S) applications. Contractors outside of DHS typically develop homeland security M&S applications. Often the M&S applications that are developed today are difficult to extend for new purposes, to integrate with other software systems, or to be used by persons other than the original developer or development team. Slight changes to a model or scenario may require major changes to a system and significant cost to the sponsor.

Examples of topics (within the context of simulation) that are addressed in this report include:

- Modeling and simulation technologies appropriate to different problem domains
- Simulation engines
- User interfaces
- Inputs and output data types
- Databases, data files, and translators
- Information security mechanisms
- Project team and developer qualifications
- Analysis of user needs and system requirements
- Simulation specification: models, designs, and interfaces
- Software risk management
- Proprietary commercial versus open source software
- Standards
- Distributed architectures and communications
- Verification, validation, accreditation, and testing
- Documentation and training

For each topic, the following information is provided: a brief introduction to the topic, explanation of its significance, and definitions of key terminology. Additional information includes discussion of potential problem area(s), relevant technologies, the possible role of standards, expectations with respect to deliverables, recommendations and guidelines, and sources of further information.

The technical guidance contained in this report is based on published literature, information available on the Internet, technical expertise, and personal experience.

This page left blank intentionally.

Table of Contents

This page left blank intentionally

1 Introduction

Models and simulations (M&S) can be extremely useful tools for solving difficult problems, understanding the behavior of complex systems, conducting disaster exercises at the local and national levels, and training personnel on organizations, policies, operations, and procedures. Unfortunately, M&S applications do not always live up to expectations. M&S applications may not produce valid or reliable results, may provide negative training, or may be difficult to adapt to changing needs. They also may be costly and time-consuming to develop, operate, and maintain. Development of M&S applications typically requires special expertise, often a team of experts.

The U.S. Department of Homeland Security (DHS) has a broad mission and a wide variety of M&S applications may ultimately be developed to support this mission. For example, M&S applications may be developed to analyze the performance of critical infrastructure systems, train incident management personnel, evaluate resource requirements for healthcare systems, or model the spread of plumes resulting from the release of hazardous materials. Within the context of the DHS M&S applications, development is typically performed by outside contractors, government laboratories, and the academic establishment.

This report provides guidance for personnel involved in the maintenance, support, and deployment of homeland security models and simulations. Personnel may have significant expertise in some areas pertaining to the development of models and simulations, but may be unaware of important issues in other areas due to scope and complexity of M&S within DHS. This document attempts to provide a high-level overview of a number of key areas pertaining to homeland security M&S. It is intended to provide useful information for a broad range of topics that may help personnel better understand areas that they may not be familiar with, raise key issues, and answer some important questions with respect to those areas.

For the purpose of the scope of this document, M&S applications are limited by information technology and software systems that are primarily used for analysis, management planning, or training needs. Although concepts addressed in this document should be applicable to most DHS M&S domains, it is only intended to focus on four application domains where M&S is currently being applied: critical infrastructure systems, incident management systems, healthcare systems, and hazardous material releases.

1.1 Purpose

This document was prepared to assist personnel who may not be experts in simulation technology to better understand M&S capabilities and issues associated with the specification, development, and operation of M&S applications. Information contained in this report may be useful for the creation of checklists of points to be addressed as new simulation applications are created for DHS. The report is intended to help establish a common ground between DHS personnel, external developers, and others in the M&S community. Several topics associated with the archiving, configuration management, and deployment of M&S applications after they are delivered to DHS are the subject of a separate report, see [NIST 2012].

The intended audience for this report is DHS program/project managers, support personnel, contracting officers technical representatives (COTRs), external contractors and researchers, and users of M&S technology within and outside of DHS. The document is being made publicly available through the NIST Publications Portal [NIST 2011a] and is accessible online to all who have interest in this area.

Some of the topics that are addressed in this report are: 1) the scope of homeland security domains for which the information contained within applies, 2) technologies that may be employed to develop M&S applications, 3) key components of M&S applications, and 4) the simulation development process.

For each topic, an introduction and background discussion, key terminology, issues, recommendations and guidance, and pointers to further information are also provided.

1.2 Background of M&S Policy and Guidance

Congressional leaders formed the Congressional M&S Caucus to "…showcase M&S initiatives, promote the M&S industry, and …be a forum to understand the policy challenges facing this growing and versatile technology." In 2007, the Caucus sponsored House Resolution 487, which recognizes the contribution of M&S technology to the security and prosperity of the United States. The resolution includes M&S as a National Critical Technology and recognizes its role in planning for national disasters and emergency preparedness response as below [HR 2007].

> …. *Modeling and simulation efforts have, and will continue to –*
> - *Provide vital strategic support functions to our Military;*
> - *Defend our freedom and advance United States interests around the world;*
> - *Encourage comprehensive planning for national disaster and emergency preparedness response;*
> - *Improve and secure our critical infrastructure and transportation systems.*

In 2012, a new voluntary non-profit national organization is forming called the National Modeling and Simulation Coalition (NMSC). The intent of this organization is to bring together all constituents and stakeholders using M&S in healthcare, energy, education, defense, homeland security, engineering, manufacturing, transportation and other areas. Its purpose is to establish and further a national agenda for M&S and to drive its growth and use into broad areas to support our national economy, welfare, and security [NDIA 2012].

Federal laws, policies, and strategies include M&S capabilities for policy analysis, regulatory environmental standards, national defense, homeland security, and a variety of other purposes. Those related to homeland security include:

- The Homeland Security Act [DHS 2002a].
- The Patriot Act [Patriot 2001].
- The National Strategy for Homeland Security [DHS 2007].
- The National Infrastructure Protection Plan [NIPP 2009].
- Homeland Security Presidential Directives (HSPD) [HSPD 2011].
- The 9/11 Commission Act [Senate 2007].

Legislation established several major M&S centers to support the DHS mission:

- The National Infrastructure Simulation and Analysis Center (NISAC) supports the Office of Infrastructure Protection and provides modeling and analysis for the National Infrastructure Protection Plan (NIPP).
- The Interagency Modeling and Atmospheric Assessment Center (IMAAC) supports the National Response Framework (NRF) [FEMA 2008] and serves as the single Federal source of airborne hazards predictions during an Incident of National Significance (INS).

- A National Exercise Simulation Center has been established by the Federal Emergency Management Agency (FEMA). The center will use M&S for training, exercises, and command and control functions at the operational level. These M&S capabilities will include a variety of live, virtual, and constructive simulations to prepare elected officials, emergency managers, emergency response providers, and emergency support providers at all levels of government to operate cohesively.

The Homeland Security Strategy of 2002 [DHS 2002b] stated that DHS will "Develop the best modeling and simulation tools to understand how our increasingly complex and connected infrastructures behave, and to shape effective protection and response options." The National Infrastructure Protection Plan [NIPP 2009] repeatedly cites the central role of M&S, stating that, "Additional DHS critical infrastructure/key resource (CI/KR) protection roles and responsibilities include conducting modeling and simulations to analyze sector, cross-sector, and regional dependencies."

HSPD-7 [HSPD 2011] mandates that, "The Secretary will utilize existing, and develop new, capabilities as needed to model comprehensively the potential implications of terrorist exploitation of vulnerabilities in critical infrastructure and key resources." HSPD-18 and HSPD-21 further cite the need for high-level modeling of weapons of mass destruction (WMD) scenarios and model-based assessment of public health preparedness.

Federal legislation and policies mandate that DHS coordinate development and use of M&S capabilities with other Federal agency partners and the private sector. Federal legislation and policies also highlight the importance of M&S applications, especially for assessing risks to critical infrastructure and responding to damage from natural and man-made disasters and catastrophes.

From a federal information technology (IT) perspective, [EOP 2007] includes M&S in the business analytical services domain as knowledge discovery capabilities, which facilitate the identification of useful information from data:

Modeling: Develop descriptions to adequately explain relevant data for the purpose of prediction, pattern detection, exploration or general organization of data.
Simulation: Utilize models to mimic real-world processes.

Many Federal organizations use M&S and have established programs, policies, organizations, guidelines, and/or standards to support the management, coordination, development, evaluation, and use of these capabilities. [Hutchings 2010] summarizes M&S management for the U. S. Department of Defense (DoD), U. S. Department of Energy (DOE), U. S. Environmental Protection Agency (EPA), and National Aeronautics and Space Administration (NASA).

DoD has used M&S capabilities for decades to support training, analysis, and acquisition for defense at many levels. The Defense Modeling and Simulation Office (DMSO) was established in 1991 to foster joint interoperability and reuse among Service M&S efforts, based on Congressional interest and a simulation policy study [Hollenbach 2009]. An Executive Council for Modeling and Simulation (EXCIMS) was established concurrently to improve oversight and coordination of M&S across DoD.

DoD has also established a Strategic Vision for Modeling and Simulation that states the following [DoD 2007a]:

> *Empower DoD with modeling and simulation capabilities that effectively and efficiently support the full spectrum of the Department's activities and operations.*

End State: A robust modeling and simulation (M&S) capability enables the Department to more effectively meet its operational and support objectives across the diverse activities of the services, combatant commands, and agencies. A defense-wide M&S management process encourages collaboration and facilitates the sharing of data across DoD components, while promoting interactions between DoD and other government agencies, international partners, industry, and academia.

The goals of DoD's M&S efforts are to provide:

1. Standards, architectures, networks and environments that:
  - *Promote the sharing of tools, data, and information across the Enterprise.*
  - *Foster common formats.*
  - *Are readily accessible and can be reliably applied by users.*

2. Policies at the enterprise level that:
  - *Promote interoperability and the use of common M&S capabilities.*
  - *Minimize duplication and encourage reuse of M&S capabilities.*
  - *Encourage research and development to respond to emerging challenges.*
  - *Limit the use of models and data encumbered by proprietary restrictions.*
  - *Leverage M&S capabilities across DoD, other government agencies, international partners, industry, and academia.*

3. Management processes for models, simulations, and data that:
  - *Enable M&S users and developers to easily discover and share M&S capabilities and provide incentives for their use.*
  - *Facilitate the cost-effective and efficient development and use of M&S systems and capabilities.*
  - *Include practical validation, verification, and accreditation guidelines that vary by application area.*

4. Tools in the form of models, simulations, and authoritative data that:
  - *Support the full range of DoD interests.*
  - *Provide timely and credible results.*
  - *Make capabilities, limitations, and assumptions easily visible.*
  - *Are useable across communities.*

5.   People that:
  - *Are well-trained.*
  - *Employ existing models, simulation, and data to support departmental objectives.*
  - *Advance M&S to support emerging departmental challenges.*

Current DoD policy on M&S specifies [DoD 2007b]:

- Establishing a management and administrative structure for improving the oversight, coordination, and communication of DoD M&S issues.
- Developing a DoD M&S Master Plan.
- Developing a coordinated DoD M&S Investment Plan.
- Investments to promote enhancements of DoD M&S technologies in support of operational needs and the acquisition process; develop common tools, methods, and databases; and establish standards and protocols promoting the use of the internet, data exchange, open systems architecture, and software reusability of M&S applications.
- Establishing an M&S Analysis Center.

M&S applications used to support major DoD decision-making organizations and processes shall be accredited for use by the DoD Component for its own forces and capabilities. The DoD Modeling and Simulation Coordination Office (MSCO), which evolved from DMSO, is also assigned responsibilities for:

- Encouraging improved communication and coordination among DoD M&S activities.
- Functioning as the DoD focal point for M&S and ensuring that M&S technology development is consistent with other related initiatives.
- Serving as Secretary to the DoD Executive Council for Modeling and Simulation.
- Chairing the DoD Modeling and Simulation Working Group (MSWG) and monitoring the activities of M&S-related joint, functional, sub working groups, and task forces.
- Staffing and distributing DoD M&S plans, programs, policies, procedures, and publications.

The DoD Components are also assigned responsibilities, e.g., for establishing verification, validation, and accreditation (VV&A) policies and procedures for the M&S applications that they manage. Each DoD Component has organizations devoted to M&S management.

The National Nuclear Security Administration (NNSA) in DOE manages the Stockpile Stewardship Program (SSP), which is a single, highly integrated technical program for maintaining the safety and reliability of the U.S. nuclear stockpile [NNSA 2004]. A cornerstone of SSP is the Advanced Simulation and Computing Program (ASC) that provides simulation capabilities and computational resources to support the annual stockpile assessment and certification, study advanced nuclear-weapons design and manufacturing processes, analyze accident scenarios and aging weapons, and provide the tools to enable Stockpile Life Extension Programs. The major sub-programs of ASC include:

- *Integrated Codes* - constitutes lab software code projects that develop and improve the weapons simulation tools, the physics, the engineering, and the specialized codes.
- *Physics and Engineering Models* - develops microscopic and macroscopic models of physics and material properties, as well as improved numerical approximations to the simulation of transport for particles and x-rays and other critical phenomena.
- *Verification and Validation* - establishes a technically rigorous foundation for the credibility of code results based on the functional and operational requirements established by designers, analysts, and code developers to improve the fidelity of codes and models.
- *Computational Systems and Software Environment* - provides ASC users a stable, seamless computing environment for all ASC-deployed platforms, including capability, capacity, and advanced systems.
- *Facility Operations and User Support* - provides both necessary physical facility and operational support for reliable production computing and storage environments as well as a suite of user services for effective use of ASC tri-lab computing resources. The scope of the facility operations includes planning, integration and deployment; continuing product support; software license and maintenance fees; procurement of operational equipment and media; quality and reliability activities; and collaborations.

ASC is primarily implemented in the DoE national laboratories, especially Lawrence Livermore, Los Alamos, and Sandia National Laboratories. The ASC interfaces with other government agencies such as DoD and has a strategic planning process to enhance capabilities.

EPA has the mission to protect human health and to safeguard the natural environment – air, water, and land – upon which life depends. EPA uses a variety of models to carry out this mission, including models for atmospheric and indoor air quality, chemical equilibrium, leaching and runoff, risk assessment, ground water, surface water, and toxico-kinetic effects. For example, guidelines for air quality models are specified in law, e.g., 40 Code of Federal Regulation Part 51. EPA has developed agency guidelines to support the development, evaluation, and use of computational models and simulations [EPA 2002], [EPA 2009]. An EPA guideline approved in March 2009 states:

> ...aims to facilitate a widespread understanding of the processes for model development, evaluation, and application and thereby promote their appropriate application to support informed decision making. Recognizing the diversity of modeling applications throughout the Agency, the principles and practices described in the guidance apply generally to all models used to inform Agency decisions, regardless of domain, mode, conceptual basis, form, or rigor level (i.e., varying from screening-level applications to complex analyses)...

> This document is intended for a wide range of audiences, including model developers, computer programmers, model users, policy makers who work with models, and affected stakeholders. Model users include those who generate model output (i.e., who set up, parameterize, and run models) and managers who use model outputs. [EPA 2009]

NASA approved an agency-wide standard [NASA 2008] for M&S. The intent of this standard is:

> To provide uniform engineering and technical requirements for processes, procedures, practices, and methods that have been endorsed as [a] standard for models and simulations developed and used in NASA programs and projects, including requirements for selection, application, and design criteria of an item. [NASA 2008]

This standard addresses a number of concerns that NASA has about using M&S capabilities to:

- Identify best practices to ensure that knowledge of operations is captured in the user interfaces (e.g., users are not able to enter parameters that are out of bounds).
- Develop process for tool verification and validation, certification, re-verification, re-validation, and re-certification based on operational data and trending.
- Develop standard for documentation, configuration management, and quality assurance.
- Identify any training or certification requirements to ensure proper operational capabilities.
- Provide a plan for tool management, maintenance, and obsolescence consistent with modeling/simulation environments and the aging or changing of the modeled platform or system.
- Develop a process for user feedback when results appear unrealistic or defy explanation.
- Include a standard method to assess the credibility of the models and simulations presented to the decision maker when making critical decisions (i.e., decisions that affect human safety or mission success) using results from models and simulations.
- Assure that the credibility of models and simulations meet the project requirements.

Other Federal organizations are using M&S capabilities that are managed as programs supporting specialized communities or key functional areas. Some of these communities have developed guidelines for evaluation and use of M&S capabilities in their subject areas, e.g., see [GAO 1976], [GAO 1978], [GAO 1979], [Cloyd 2007], and [Neuman 2002].

1.3 Background of M&S Implementation within DHS

Modeling and simulation activities are being actively pursued within DHS and by DHS contractors. DHS M&S activities are constantly evolving. This section introduces some topics pertaining to M&S within DHS, namely: 1) component organizations involved in M&S and 2) roles and responsibilities of DHS staff that potentially have a stake in M&S.

1.3.1 DHS Component Organizations Involved in M&S

DHS is already making widespread use of M&S. This section identifies some of the organizations within DHS that are currently using M&S and how they are using it. M&S capabilities are being used to support a number of different types of risk assessments, analyses, training, exercises, and system engineering needs. Analytical capabilities supporting problem solving and decision making include:

- Analysts and analytical expertise.
- Analytical methods and processes.
- Body of knowledge (data, information, reports, etc.).
- Tools (models, simulations, computational capabilities, experiments, etc.).

For example, the Office of Infrastructure Protection (OIP) Infrastructure Analysis and Strategy Division (IASD) conducts risk analysis supported by contractor analysts and dedicated personnel at DoE national laboratories. IASD manages multiple models/simulations and data sets through a capability portfolio. IASD, in coordination with the Infrastructure Information Collection Division (IICD), acquires a wide variety of both government and commercial data.

A list of M&S capabilities currently used by DHS and an overview of these capabilities are listed in the appendix. Some examples of modeling and simulation tools used in DHS to support M&S are shown in Table 1. The United States Coast Guard (USCG) has compiled a comprehensive list of USCG analytical tools and capabilities that are included in a Modeling and Simulation Resource Repository (MSRR). Please see the report on the DHS-NIST 2011 workshop on Modeling and Simulation Applications for Homeland Security for more homeland security M&S applications [NIST 2011b].

Table 1: Examples of DHS M&S Tools

Name	Overview
Coast Guard Maritime Operational Effectiveness Simulation (CGMOES)	CGMOES is a multi-mission discrete event simulation developed using the Maritime Operations Simulation Tool. It simulates the core functionality of Coast Guard maritime operations reflecting policy, doctrine, concept of operations (CONOPS), and logistics impacts. It is designed to support performance assessments and alternatives analysis at the campaign level, and supports measurement of projected system performance improvements.
Hazards U.S. Multi-Hazard (HAZUS-MH)	HAZUS-MH is a nationally applicable software program and standardized methodology for estimating potential losses from earthquake, flood, and hurricane hazards. FEMA developed HAZUS-MH in partnership with the National Institute of Building Sciences (NIBS). Loss estimates produced with HAZUS-MH are based on current scientific and engineering knowledge regarding the effects of earthquake, flood, and hurricane hazards.

1.3.2 DHS Roles and Responsibilities Relevant to M&S

There are a number of positions in different disciplines within DHS that potentially have a stake in the development of homeland security M&S applications. The roles and responsibilities of organization and position types are identified below. [DHS 2011] defines the core team in the DHS Program Management Office (PMO). Identified career fields and related responsibilities include:

- *Program Management:* Acquisition professionals in the program management acquisition discipline are concerned with all of the functions of a PMO. Program management professionals serve in a wide range of PMO and component acquisition staff positions, including program integrators and analysts, program managers, and their deputies. They may also serve in a number of support and management positions throughout the workforce.
- *Systems Engineering*: Systems engineers demonstrate how systems engineering technical and technical management processes apply to acquisition programs; interact with program Integrated Product Teams (IPTs) regarding the proper application of systems engineering processes; develop system models and work breakdown structures; and use top-down design and bottom-up product realization.
- *Acquisition/Financial Management:* Acquisition/Financial Management staff plan, direct, monitor, organize, and control financial resources including: formulation of budget to requirements, execution, financial systems, appropriations-related congressional issues, and reporting. They are responsible for all financial and budgeting for the program.
- *Life Cycle Logistics (LCL) Management:* LCL management is the planning, development, implementation, and management of a comprehensive, affordable, and effective systems support strategy. LCL encompasses the entire system's life cycle including acquisition (design, development, test, produce, and deploy), sustainment (operations and support), and disposal. Life-cycle logisticians perform a principal joint and component logistics role during the acquisition, operational, and disposal phases of the system's life cycle.
- *Test and Evaluation (T&E):* T&E Managers are engineers, scientists, operations research analysts, system analysts, computer scientists, and other technical personnel who plan, perform, and manage T&E tasks in support of all acquisitions. The individuals in T&E positions are subject matter experts who will plan, monitor, manage, and conduct T&E of prototype, new, fielded, or modified IT; non-IT; Command, Control, Communications, Computers, Intelligence, Surveillance, and Reconnaissance (C4ISR); and infrastructure systems.
- *Cost Analysts*: Cost analysts lead teams in the analysis of schedule requirements, gathering data relevant to the system-to-be cost, development or evaluation of existing cost estimating relationships and mathematical models, performance of risk analysis on program assumptions, analysis of results of models and other methods of developing costs, accumulation of cost estimates made by work breakdown structure elements, and development of summary data for presentation purposes; development in written, graphical and tabular from information on methods and data used during the development of the estimate, reconciliations of program content and methods used by the organization-independent estimate vice those used by the program office of the weapon/information system being costed; preparation of briefings to management in development of cost databases, parametric relationships, cost estimating software and documentation, and research reports.
- *Information Technology (IT)*: IT specialists include computer scientists, information technology management specialists, computer engineers, and telecommunications managers, who directly support the acquisition of information technology. This may include hardware, software, or firmware products used to create, record, produce, store, retrieve, process, transmit, disseminate, present, or display data or information.

- *Contracting Officer's Technical Representative (COTR)*: COTR is a business communications liaison between the United States Government and a private contractor. COTRs ensure that their goals are mutually beneficial. The COTR is responsible for recommending and authorizing (or denying) actions and expenditures for both standard delivery orders and task orders, and those that fall outside of the normal business practices of its supporting contractors and sub-contractors.

1.4 Document Overview

Section 2 describes the homeland security domains for which the information contained within this document applies. Section 3 identifies simulation technologies that are applicable to the homeland security M&S applications for the domains that were just identified as well as development technologies that may be used to create those applications. Section 4 describes component modules that may typically comprise simulation applications. Section 5 presents an overview of the M&S application development process. Section 6 presents document conclusions. Section 7 provides definitions for selected abbreviations and acronyms that appeared in this report. Section 8 identifies references and sources of further information on the topics that were addressed.

2 Homeland Security M&S Application Domains

This section provides a brief summary of four homeland security M&S domains that are the primary focus of this report. It also identifies some modeling domains that are specifically not addressed, although information contained herein may possibly be relevant to those domains. The four domains that are addressed include critical infrastructure systems (CI), incident management (IM), hazardous material releases (HMR), and healthcare systems (HS).

2.1.1 Critical Infrastructure (CI) Systems

Introduction – Protecting and ensuring the continuity of critical infrastructure systems are essential to the nation's security, public health and safety, economic vitality, and way of life. Eighteen critical infrastructures and key resources have been defined in the National Infrastructure Protection Plan [NIPP 2009]: Agriculture and Food, Banking and Finance, Chemical, Commercial Facilities, Communications, Critical Manufacturing, Dams, Defense Industrial Base, Emergency Services, Energy, Government Facilities, Healthcare and Public Health, Information Technology (IT), National Monuments and Icons, Nuclear Reactors, Materials, and Waste, Postal and Shipping, Transportation Systems, and Water.

To ensure an effective, efficient Critical Infrastructure and Key Resource (CIKR or CI/KR) protection program over the long term, the NIPP relies on the development, safeguarding, and maintenance of data systems and simulations to enable continuously refined risk assessment within and across sectors and to ensure preparedness for domestic incident management. In particular, the complexity of interdependency calls for use of modeling and simulation capabilities. The DHS Office of Infrastructure Protection (IP) is the lead coordinator for modeling and simulation capabilities regarding CIKR protection and resiliency [NIPP 2009]. In this capacity, DHS has several objectives including:

- Specifying requirements for the development, maintenance, and application of research and operations-related modeling capabilities for CIKR protection and resiliency.
- Utilizing available relevant modeling and simulation capabilities in training and exercises to familiarize the Sector Specific Agencies (SSAs) and other CIKR partners with them.
- Providing guidance on the vetting of modeling tools.
- Promoting use of applicable private sector modeling capabilities, including initiatives and expertise, by DHS, SSAs, and their CIKR partners.

The principal capability within the Infrastructure Protection (IP) component to support modeling, simulation, and analysis efforts is the National Infrastructure Simulation and Analysis Center (NISAC). NISAC has been tasked with developing advanced modeling, simulation, and analysis capabilities for the Nation's CIKR. These tools and analyses together address CIKR in all-hazards context including physical and cyber dependencies and interdependencies. These sophisticated models better inform decision-makers, especially for cross-sector priorities. NISAC is the principal but not the sole source available to CIKR stakeholders in need of modeling, simulation, and analysis capabilities. NISAC works with other providers of CIKR analysis to improve overall analytical quality and ensure consistency.

Information regarding sector-specific CIKR-related authorities is addressed in the respective Sector Specific Plans (SSPs). The SSPs provide the means for implementing NIPP across all sectors. SSPs also provide a national framework for each sector that guides the development, implementation, and updating of State and local homeland security strategies and CIKR protection programs [NIPP 2009]. As the responsible agent for the identification of assets and existing databases for their sectors, the SSAs' objectives include:

- Outlining the sector plans and processes for database, data system, and modeling and simulation development and updates in their respective SSPs.
- Facilitating the collection and protection of accurate information for database, data system, and modeling and simulation use in collaboration with sector partners.

Fourteen of the SSPs are publicly available. The limited current capabilities identified in these SSPs suggest that there are large opportunities for use of M&S in each of the sectors.

CIKR models and simulations may be used to understand infrastructure systems, their interdependencies, their vulnerabilities, and the impact of the propagation of damage across interdependent infrastructure systems based upon emergency incidents. They may also be used to support training exercises, performance measurement, conceptual design, impact evaluation, response planning, analysis, acquisition, conceptualizing and evaluating new systems, vulnerability analysis, risk analysis, economic impact, and determining interdependencies between CIKR systems.

Terminology:

> **Critical Infrastructures** – the assets, systems, and networks, whether physical or virtual, so vital to the nation that their incapacitation or destruction would have a debilitating effect on national economic security, public health and safety, or any combination thereof.

> **Key Resources** – publicly or privately controlled resources essential to the minimal operations of the economy and government.

> **Vulnerability Analysis** – a systematic examination of a system or product to determine the adequacy of security measures, identify security deficiencies, provide data from which to predict the effectiveness of proposed security measures, and confirm the adequacy of such measures after implementation [Free 2011].

> **Risk Analysis** – a technique to identify factors and probabilities associated with those factors that affect a project's success or the achievement of a goal.

> **Performance Measurement** - The National Performance Review provides a complimentary definition of performance measurement:

> > *A process of assessing progress toward achieving predetermined goals, including information on the efficiency with which resources are transformed into goods and services (outputs), the quality of those outputs (how well they are delivered to clients and the extent to which clients are satisfied) and outcomes (the results of a program activity compared to its intended purpose), and the effectiveness of government operations in terms of their specific contributions to program objectives.* [DoT 2011]

Issues:

> **Potential Problem Area(s)** – This M&S domain represents a vast and diverse array of critical infrastructures, facilities, systems, information, personnel, and other resources. The development of detailed, accurate, validated models and simulations of each local infrastructure instance is a task of almost unimaginable magnitude. The National Infrastructure Simulation and Analysis Center (NISAC) Fast Analysis and Simulation Team (FAST) has developed a number of CIKR models and simulations. These M&S provide insights into the interdependency and economic impacts of disruptions to infrastructure elements that might result from an event or set of events, a

policy change, or implementation of protective measures. [NISAC 2011b]. Unfortunately, these models and simulations can typically provide general insights at a fairly high level. Ultimately, infrastructure and key resource organizations/managers would benefit from M&S tools that can be used to model interdependencies at the local level. New M&S tools and standard data structures will need to be implemented to make this technology accessible at a local level.

Relevant Technologies – All of the modeling and simulation technologies addressed in subsequent sections of this report are relevant to CIKR systems, e.g., system dynamics, discrete event, agent based, and physical-science-based simulations.

Possible Role of Standards – Standard reference models that define the systems, functions and data for each critical infrastructure are needed. These reference models do not currently exist. The Unified Modeling Language (UML) provides a powerful set of tools for creating such reference models [OMG 2012].

Expectations With Respect To Deliverables – M&S applications developed for CIKR systems should contain appropriate documentation, e.g., needs analyses, requirements specifications, reference models, software design information, M&S validation data, and user manuals.

Recommendations and Guidance – As a first step towards establishing a consistent approach to M&S for CIKR, generic reference models should be created. These models could be generated as a part of ongoing projects. Models are needed that identify:

- For each critical infrastructure area: component facilities and systems and data, interactions/interdependencies between infrastructure components and other areas, internal facilities and systems, and outside factors affecting its mission and operations.
- Physical and information security mechanisms for facilities and systems.

A repository of CIKR reference models should be established for use by future development efforts.

For Further Information – For more details on critical infrastructure systems, see the National Infrastructure Protection Plan [NIPP 2009]. For more on the NISAC FAST, see [NISAC 2011b].

2.1.2   Incident Management (IM)

Introduction – The National Incident Management System (NIMS) [DHS 2008] and National Response Framework (NRF) [FEMA 2008] provide the relevant guidance for the incident management domain. The NRF defines the roles and responsibilities, response actions, and organizations, and emphasizes planning as a critical element of the response. It introduces the National Planning Scenarios (NPS) to be used as a critical element for preparedness.

In the National Response Framework, DHS defines 15 emergency support functions (ESFs) [FEMA 2011]. The functions define a common set of tasks that organizations within and outside of government perform to prepare for and respond to natural disasters, terrorist attacks, industrial accidents, and other incidents. The functions that have currently been defined are:

1. *Transportation* processes and coordinates requests for Federal and civil transportation support, reports damage to transportation infrastructure as a result of the incident, coordinates alternate transportation services and the restoration and recovery of the transportation infrastructure, and coordinates and supports prevention/preparedness/mitigation among transportation infrastructure

stakeholders at the State and local levels.

2. *Communications* provides the required temporary telecommunications and helps restore telecommunications infrastructure.

3. *Public Works and Engineering* conducts pre- and post-incident assessments of public works and infrastructure, executes emergency contract support of life-saving and life-sustaining services, provides technical assistance such as engineering expertise, provides emergency repair of damaged infrastructure, and implements and manages Public Assistance Program and other recovery programs.

4. *Firefighting* focuses on the detection and suppression of wild land, rural, and urban fires and provides personnel, equipment, and supplies in support of the firefighting operations.

5. *Emergency Management* facilitates information flow in the pre-incident prevention, supports and facilitates multi-agency planning and coordination during the post-incident phase.

6. *Mass Care, Emergency Assistance, Housing, and Human Services* provides economic assistance and other services for individuals, households, and families impacted by the incident; the services may include sheltering of victims, organizing feeding operations, providing emergency first aid, providing short- and long-term housing needs, and providing victim-related recovery efforts such as counseling.

7. *Logistics Management and Resource Support* provides emergency relief supplies, facility space, office equipment, office supplies, telecommunications, contracting services, transportation services, security services, and personnel for immediate response activities.

8. *Public Health and Medical Services* provides supplemental assistance for the public health and medical needs of victims.

9. *Search and Rescue* provides specialized life-saving assistance including locating, extricating, and providing onsite medical treatment to victims trapped in collapsed structures.

10. *Oil and Hazardous Materials Response* provides the hazard-specific response mechanisms to support actual or potential oil and hazardous materials incidents.

11. *Agriculture and Natural Resources* provides nutrition assistance; animal and plant disease and pest response; safety and security of the commercial food supply; and protection of natural resources, cultural resources, and historic properties.

12. *Energy* is designed to support incident management requirements including force and critical infrastructure protection, security planning and technical assistance, technology support, and public safety in both pre-incident and post-incident situations.

13. *Public Safety and Security* provides a mechanism for coordinating and providing supports that include non-investigative/non-criminal law enforcement, public safety, and security capabilities and resources.

14. *Long-Term Community Recovery* provides a framework for Federal Government support, including Federal's available programs and resources, to enable community recovery from the long-term consequences of the incident and to reduce or eliminate risk from future incidents.

15. *External Affairs* coordinates Federal actions to provide the resource support and mechanisms to

Federal, State, local, and tribal incident management elements to ensure the required public affairs support or assets are employed.

The NRF also defines 15 national planning scenarios [FEMA 2009] that may be used to help focus efforts to prepare for natural disasters, terrorist attacks, and other serious incidents. The national planning scenarios are Improvised Nuclear Device; Aerosol Anthrax; Pandemic Influenza; Plague; Blister Agent; Toxic Industrial Chemicals; Nerve Agent; Chlorine Tank Explosion; Major Earthquake; Major Hurricane; Radiological Dispersal Device; Improvised Explosive Device (IED); Food Contamination; Foreign Animal Disease; and Cyber Attack. See [FEMA 2009] for further information on the national planning scenarios.

Incident management models and simulations may be used to support analysis, planning, and training needs pertaining to terrorist attacks, national security events, and natural and man-made disasters. Simulation models may be used to understand incident management systems, interdependencies with other systems, their vulnerabilities, and the impact of emergency incidents on the population and responder community. Incident management models and simulations will be used to support training exercises, performance measurement, conceptual design, impact evaluation, response planning, analysis, acquisition, conceptualizing and evaluating new systems, vulnerability analysis, economic impact, and determining interdependencies between incident management and other infrastructure systems.

Terminology: Incident and incident management, as defined by NIMS appears below:

Incident – "An occurrence, natural or manmade, that requires a response to protect life or property. Incidents can, for example, include major disasters, emergencies, terrorist attacks, terrorist threats, civil unrest, wild land and urban fires, floods, hazardous materials spills, nuclear accidents, aircraft accidents, earthquakes, hurricanes, tornadoes, tropical storms, tsunamis, war-related disasters, public health and medical emergencies, and other occurrences requiring an emergency response." [DHS 2008]

Incident Management – "The broad spectrum of activities and organizations providing effective and efficient operations, coordination, and support applied at all levels of government, utilizing both governmental and non-governmental resources to plan for, respond to, and recover from an incident, regardless of cause, size, or complexity." [DHS2008]

Issues: Technical issues that are discussed below include identification of applicable simulation technologies, problems affecting the development of M&S applications, expectations for delivered products, and the possible role of standards in the development of IM M&S applications:

Potential Problem Area(s) – The diversity of incident management systems across the nation is a potential problem. It would be beneficial and cost effective to have a general set of M&S applications that could be used by many IM organizations nationwide. These M&S tools could be used to analyze the behavior and performance of organizations and systems, plan responses to different types of incidents, and train personnel on response procedures. Unfortunately, individual incident management organizations have a wide variety of different names, responsibilities, resources, databases, and file formats across the country. This diversity complicates the development of models and simulations that can be widely used. For example, a simulation that is developed for one organization may not accept data used by another organization without the development of special purpose data translators. Depending upon the data involved, it may not be possible to translate one organization's data into the form required by a particular simulation.

Relevant Technologies – Within the context of incident management, several of the simulation technologies that are introduced in Section 3.1 may be found to be applicable. System dynamics simulations may be used to develop high-level representations of incident management system components and their interactions. Discrete event simulations may be used to better understand information and material flows within the National Incident Management System (NIMS) organizational structure, resource requirements and utilization, schedules, and response times to different types of incidents. Gaming systems and tabletop enablers may be used to support individual and team training exercises for incident managers and first responders.

Possible Role of Standards – Incident management M&S applications should utilize appropriate geographic information system (GIS), emergency communications messaging, and symbology standards.

Expectations With Respect To Deliverables – Incident management M&S applications should adhere to the structure of the NIMS and be compatible with the ESFs. Wherever possible forethought should be given to the adaptation of simulation to different IM scenarios to enable M&S reuse. M&S documentation should clearly explain how to adapt M&S software to new or modified scenarios.

Recommendations and Guidance – A set of hierarchical incident management reference models could help facilitate the development of compatible homeland security simulations. Reference models should be developed using the Unified Modeling Language (UML) and integrate elements of NIMS with the 15 ESFs. The reference models would serve as a common basis for the development of incident management simulations by the M&S community. The models would help the M&S community to develop compatible simulations and help enable the integration of those simulations. See Section 5.3 for a more detailed discussion about modeling with UML.

From the vast array of standards that may be used to support Incident Management, e.g., geographical information system (GIS) standards, DHS should work with the M&S and responder communities to identify a preferred set of standards for applications development. If there are gaps in the standards that are needed, DHS should help promote new consensus standards efforts (see also Section 5.6, Use of Standards).

For Further Information – See [DHS 2008] for further information on the National Incident Management System, [FEMA 2008] for the Framework (NRF), and [FEMA 2009] for further information on the 15 National Planning Scenarios.

2.1.3   Hazardous Material Releases (HMR)

Introduction – Hazardous materials (HAZMAT) are substances that if released or misused can cause death, serious injury, long lasting health effects, and damage to structure and other properties as well as to the environment [GBRA 2010]. Hazardous materials may be in solid, liquid, or gaseous form and may be explosive, flammable, combustible, corrosive, reactive, poisonous, biological, or radioactive. These materials have to be properly contained in storage, use and transport, else their chemical, physical, and biological properties may pose a potential risk to life, health, the environment, and property. In a hazardous materials incident, solid, liquid, or gaseous contaminants may be released from fixed or mobile containers. Hazardous material incidents can range from an accident on the highway resulting in a chemical spill to contamination of groundwater by naturally occurring methane gas.

There are numerous incidents of hazardous material releases (HMR) in the United States every year. The releases may be airborne, or spills and discharges that contaminate water bodies, vegetation, soil, and built-up structures. The airborne hazards are also identified as hazardous fumes, noxious chemicals, or mysterious odors. The airborne hazards affect areas and people outdoors but they may permeate buildings and affect people indoors. The hazardous material releases lead to areas and buildings being evacuated in a majority of cases. However, depending on the hazard, release pattern, weather conditions, and a number of other factors, it may be advisable to stay indoors, i.e., shelter in place.

A majority of the HMR incidents are results of accidental releases of toxic industrial or agricultural chemicals [AFCESA 2001]. Occasionally they may be releases of biological or radiological materials. The majority of releases happen during the course of regular operations at fixed facilities [KCOEM 2011], that is, due to industrial accidents (e.g., fire or equipment malfunction at chemical plants). Other causes of releases are transportation accidents (e.g., tanker collisions and train derailments), malicious acts (e.g., vandalism and terrorist attacks), and natural disasters (e.g., earthquakes and hurricanes). A recent example of HMR due to a natural disaster is the release of radioactive plumes from a nuclear power plant affected by an earthquake and tsunami in Japan in March 2011.

Simulations of hazardous material releases attempt to model and evaluate the dispersion of various kinds of materials including chemical, biological, nuclear, and radiological agents into the atmosphere, HVAC (heating, ventilation, and air conditioning) systems within buildings and other enclosed spaces, watershed systems, and the soil. Releases may be accidental (e.g., a ruptured tank car from a train derailment), intentional (e.g., a terrorist attack), or natural (e.g., a wildfire or volcanic eruption). Examples of release sources include nuclear power plants, storage tanks or industrial plants, chemical and biological sprayers, fires, smokestack emissions, nuclear detonation clouds, and other explosive blasts.

The airborne HMR have drawn more attention due to their potential to negatively affect large areas. Some key parameters that are relevant to dispersion of atmospheric releases are the release location, release mechanism, agent chemical/material properties, weather conditions, and terrain and geography. M&S tools are important to analyze and predict the dispersion of such releases using inputs on these parameters based on reports, visual observations, and sensors. The tools take into account the material released, the local topography, and meteorological and atmospheric data to determine the dispersion area and associated concentrations [NRC 2003]. The concentrations are then used to assess the risk to the population, environment, and property in the affected areas. Incident management personnel use the results of these types of models and simulations to predict the impact of releases, allocate resources, and plan response operations, among other uses.

Terminology:

> **Hazardous Material** – a substance capable of causing harm to people, other living beings, property, or the environment, e.g., a variety of chemical, biological, radiological, nuclear, explosive (CBRNE) substances.

Issues:

> **Potential Problem Area(s)** – Although predictions arising out of HMR models and simulations in open terrain situations (where buildings or other obstructions are not present) may be reasonably accurate, urban environments present far more complex modeling problems. Considerable research is likely to be needed to develop reliable predictive models and simulations of HMRs in urban environments. Another problem area may be M&S of indoor releases. This fact may be due more to a lack of readily retrievable building data in an appropriate format for any building of interest than limitations of HMR modeling technology. Finally, the HMR M&S community has not yet been able to reach a consensus agreement on how HMR models and simulations should be validated.

Relevant Technologies – Physical-science-based simulations, e.g., computation fluid dynamics, are the primary technology for addressing HMR M&S issues.

Possible Role of Standards – A number of standards for M&S data inputs/outputs exist and are currently in use, e.g., geographic information system (GIS) [ISO 2007], weather [NWS 2009], air quality data [ORNL 2006].

Expectations With Respect To Deliverables – Project proposals for new HMR M&S applications should clearly identify what new capabilities they are adding to the current body of knowledge in this area and how the new models and simulations will be validated.

Recommendations and Guidance – M&S efforts in the HMR domain are probably some of the most mature within the homeland security domain, as seen in the capabilities of National Atmospheric Release Advisory Center (NARAC). But, challenges still remain. It is reasonable to expect that many of the most significant HMRs will occur in high population areas, perhaps as a result of an industrial accident or a terrorist attack. Modeling of urban environments, with complex building geometries, is still an area where research is needed.

For Further Information – The National Research Council study [NRC 2003] is an excellent resource on information pertaining to the modeling of HMRs.

2.1.4   Healthcare Systems (HS)

Introduction – The Healthcare and Public Health critical infrastructure sector consists of state and local health departments, hospitals, health clinics, mental health facilities, nursing homes, blood-supply facilities, laboratories, mortuaries, and pharmaceutical stockpiles. Organizations that may be involved in addressing sector issues as well as the development of models and simulations include: Department of Health and Human Services (HHS), Center for Disease Control (CDC), Food and Drug Administration (FDA), Department of Homeland Security, Social Security Administration (SSA), the U.S. Public Health Service (USPHS), state and local agencies, academic institutions, and research hospitals.

Healthcare System models and simulations may be used to support analysis, planning, and training needs for the healthcare institutions, epidemics, and other healthcare-related emergencies. Simulation models may be used to understand healthcare systems, interdependencies with other systems, their vulnerabilities, and the impact of emergency incidents on the population and healthcare community. These models and simulations will also be used to support training exercises, performance measurement, conceptual design, impact evaluation, response planning, analysis, acquisition, conceptualizing and evaluating new systems, vulnerability analysis, economic impact, and determining interdependencies between healthcare and other infrastructure systems. Special purpose medical mannequins may be developed and used to support simulation exercises.

The healthcare systems domain includes modeling of incident victims and existing patients affected by incidents; medical symptoms, physiological processes, and behaviors that patients may experience as the result of a natural disaster, terrorist attack, or epidemic; disease management operations and procedures; the impact of disasters, etc., on the environment as well as the food supply (pollution, contamination, etc.). Issues that may be addressed by simulation applications include analysis of policies, evaluation of options and predictions concerning the state of public health, spread of communicable diseases, policies for intervention, triage and priority scoring for surgeries or other treatments, mass prophylaxis and vaccination programs, cost-effectiveness analysis, medical emergency response, fatalities management, overall readiness of the healthcare system, hospitals and other related facilities or organizations, surge capacity, operational practices, logistic support systems, and elements of the pharmaceutical/equipment manufacturing sectors serving healthcare system needs.

Healthcare system models may be comprised of:

- Representations of medical personnel and hospitals.
- Constrained medical resources and work calendars.
- Hospital facilities (e.g., operation theatres, intensive care units, and beds.)
- Medical equipment, single use devices, and consumables.
- Administrative documents and processes, and medical procedures.
- Routing, status, and location of ambulances as well as other Emergency Medical Technician (EMT) resources and processes.
- Location, cost, status, quality, effectiveness, and dispensation of pharmaceuticals.
- Logistics of industry practices and processes (e.g., climate controlled containers, and other special needs).

Terminology:

Triage – the evaluation of, sorting, prioritization/allocation of treatment to patients and disaster victims to maximize the number of survivors and appropriately utilize limited medical resources.

Mass prophylaxis – the capability to protect the health of the population through administration of critical interventions in response to a public health emergency in order to prevent the development of disease among those who are exposed or are potentially exposed to public health threats. This capability includes the provision of appropriate follow-up and monitoring of adverse events, as well as risk communication messages to address the concerns of the public." [FHA 2011]

Medical mannequin (or manikin) – Medical simulation mannequins, models, or related artifacts are widely used in medical education. These are sometimes also referred to as virtual patients. Mannequins are life-sized dolls with simulated airways or physiological systems used in the teaching of first aid, cardiopulmonary resuscitation (CPR), and advanced airway management skills such as tracheal intubation and for human figures used in computer simulation to model the behavior of the human body [Wiki 2011a].

Issues:

Potential Problem Area(s) – Healthcare system M&S needs and system requirements for homeland security CI and IM applications are currently relatively ill defined. VV&A methods for a wide variety of HS M&S applications are yet to be defined. Given the potential impact of healthcare system readiness on the population (e.g., illnesses, injuries, and deaths) for different incidents, further investment of resources in this area could have a significant payback.

Relevant Technologies – Modeling technologies such as the Unified Modeling Language (UML) are applicable to the definition of HS reference models. Each of the different types of simulation technology that are discussed later in this document are applicable to healthcare M&S, namely, system dynamics, discrete event, agent-based, physical science-based, gaming systems, and desktop enablers. UML and Extensible Markup Language (XML) are applicable to the specification of HS data sets and data interfaces.

Possible Role of Standards – A number of standards for healthcare system data are already in place, e.g., healthcare records, emergency communications, personnel protective equipment, and health hazards. Wherever possible these standards should be used directly or indirectly through the development of data translators that can reformat data so that it is more appropriate or usable by M&S applications.

Expectations With Respect To Deliverables – HS models should be properly documented. Documentation should include identification of experts and their qualifications who participated in the development process, sources of data, standards employed, VV&A information, and user documentation that indicates how models may be modified or adapted to new scenarios.

Recommendations and Guidance – A set of hierarchical healthcare system reference models could help facilitate the development of homeland security simulations. Reference models should be developed using the Unified Modeling Language (UML), indicating organizations, systems, processes, data, etc. The reference models would provide a common basis for the development of healthcare system simulations by the M&S community. The models would help the M&S community to develop compatible simulations and help enable the integration of those simulations. See Section 5.3 for a more detailed discussion about modeling with UML.

DHS should work with other organizations in the healthcare community that were identified above to identify the types of M&S applications that are needed to support homeland security. Detailed user needs, system requirements specifications, architectures, and M&S software design documents should be developed for these applications that will help guide future software development efforts.

From the vast array of standards that may be used to support healthcare M&S application development, (e.g., geographical information system (GIS) standards, healthcare records, and communications formats) DHS should work with the healthcare and M&S communities to identify a preferred set of standards for applications development. If there are gaps in the standards that are needed, DHS should help promote new consensus standards efforts (see also Section 5.6, Use of Standards).

For Further Information – For a general compendium of information on the M&S of healthcare systems including user needs, system requirements, resources, tools, standards, data sets, best practices, and issues, see [McLean 2011a]. See workshop summary of "Modeling and Simulation for Emergency Management and Health Care Systems" for further information on: 1) simulation opportunities and requirements for health care, and 2) what is needed to develop, demonstrate, and deploy a framework to enable simulations to share information within the industry [O'Hara 2010]. The appendices of [O'Hara 2010] provide related information on organizations, supporting standards associated with this area, and reference information. See [Hupert 2004] for further information on the implementation of mass prophylaxis programs.

2.1.5 Other M&S Domains

There are other domains within DHS where simulation technology may be applied but are outside the scope of this report. For example, some of these areas include support for high level DHS programming planning and budgeting, engineering and acquisition of systems, business process engineering for DHS organizations, Immigrations and Customs Enforcement activities, and Coast Guard systems and operations.

3 Technologies

This section presents a brief overview of simulation technologies (Section 3.1) and system and model development technologies (Section 3.2) that may be applicable to the homeland security modeling and simulation applications. Each simulation technology may be suitable to solving specific types of problems, but not others. The Development Technologies section identifies some of the approaches and tools that are available to develop models and simulations, irrespective of the type of simulation technology that is chosen.

3.1   Simulation Technologies

In <u>The Handbook of Simulation</u>, Jerry Banks defines simulation as:

> …the imitation of the operation of a real-world process or system over time. Simulation involves the generation of an artificial history of the system and the observation of that artificial history to draw inferences concerning the operational characteristics of the real system that is represented. Simulation is an indispensable problem-solving methodology for the solution of many real-world problems. Simulation is used to describe and analyze the behavior of a system, ask what-if questions about the real system, and aid in the design of real systems. Both existing and conceptual systems can be modeled with simulation. [Banks 1998]

DoD is a leader in the application of simulation technology to a number of different problem domains. DoD defines a simulation as a model that represents activities and interactions over time. A simulation may be fully automated (i.e., it executes without human intervention), or it may be interactive or interruptible (i.e., the user may intervene during execution). A simulation is an operating representation of selected features of real-world or hypothetical events and processes. It is conducted in accordance with known or assumed procedures and data, and with the aid of methods and equipment ranging from the simplest to the most sophisticated.

Another view of a simulation model may be defined as a representation of some or all of the properties of a device, system, or object. There are three basic classes of models: mathematical, physical, and procedural. [StrategyWorld.com 2011]

- **Mathematical model** – a representation comprised of procedures (algorithms) and mathematical equations. These models consist of a series of mathematical equations or relationships that can be discreetly solved. Usually the models employ numerical approximation techniques to solve complex mathematical functions for which specific values cannot be derived (i.e., integral).
- **Physical model** – a representation of the physical object or system and its relationship to other real-world objects.
- **Procedural model** – an expression of dynamic relationships of a situation expressed by mathematical and logical processes.

The DoD has established terminology for characterizing the interaction modes associated with different types of M&S applications, i.e., Live, Virtual, and Constructive. Live is a simulation involving real people operating real systems. A Virtual is a simulation involving real people operating simulated systems. Virtual simulations inject human-in-the-loop in a central role by exercising motor control skills (e.g., flying an airplane), decision skills (e.g., committing fire control resources to action), or communication skills (e.g., as members of a command, control, communications, computers, and intelligence team). Constructive involves simulated people operating simulated systems. Real people stimulate (make inputs to) such simulations, but are not involved in determining the outcomes. This classification has limitations as identified by DoD itself [DoD 1995]; however, it can serve as one among many dimensions for the purpose of identifying different types of M&S applications.

This report focuses on the following types of models and simulations that have widespread applicability to homeland security M&S needs:

- System dynamics.
- Discrete event.
- Agent based.
- Physical-science-based.
- Gaming systems.
- Tabletop exercise enablers.

Each of these simulation technologies is discussed below.

3.1.1 System Dynamics

Introduction – System dynamics modeling and simulation, by design, is aimed at modeling systems at high level of abstraction for supporting high level decision making. It has been applied to study a wide range of systems including industrial, social, environmental, financial, and socio-political systems, and their combinations. While generally used to model large systems at high abstraction levels, the strength of modeling feedback loops allows the technology's applications for control policies of small electro-mechanical systems.

Originally developed to study manufacturing supply chains systems (then called industrial production systems) by Professor Jay Forrester at Massachusetts Institute of Technology (MIT), system dynamics simulation is suited for studying behavior of large systems. It focuses on modeling causal relationships between key aspects of the system operating under governing policies, especially feedback loops that form beneficial or vicious cycles and determine the overall system behavior. It uses the continuous paradigm for representing time.

The technique utilizes causal loops for conceptual modeling that are enhanced into stock-and-flow diagrams for setting up the framework for computer implementation. The computer implementation then converts the causal and stock-and-flow relationships into differential equations that are used to calculate the change in system parameters over the simulated time horizon. The changes in key parameters of interest define the system performance over time.

System dynamics simulation can provide system status and changes in key parameters over time. Typical outputs are plots of system parameters over time. The plots thus provide an understanding of system behavior based on the defined structure of the system operating under defined policies and procedures. The plots of system behavior under different policy options can support selection of the option leading to desired system performance and can also help tune the policy parameters. The models are generally developed for one time use, though there are instances where system dynamics models have been encapsulated in decision support systems for use at multiple sites.

Terminology:

> **System Dynamics** – is the study of system behavior over time driven by changes in elements and structure of the processes in the system. The structure and behavior is represented using causal loop and stock-and-flow diagrams for such analysis.

Causal Loop (a.k.a. feedback loop) – is a sequence of cause and effect relationships that form a closed path and show how the factors involved lead to observed or anticipated system behavior. Each individual relationship is identified as positive (a change in one factor causes a change in the other in the same direction) or negative (change caused is in opposite direction). A number of causal loops may be used together to represent overall system behavior.

Stocks – represents accumulation of entities at any point in the process. The level of stocks may increase or decrease based on the in-flows and out-flows to the stocks. Stock may be used to represent number of ambulances, number of healthy people in population, number of infected people, etc.

Flows – represent movement of entities and are represented as rates and cause changes in the level of stocks they are associated with. Flows may be used to represent the rate at which people get infected based on the number of people that are infectious in a pandemic flu study.

Issues:

Potential Problem Area(s) – Incorrect use of system dynamics models is a potential problem. Attempts to study detailed material and information flows can lead to disappointing results. Training on effectively use of systems dynamics modeling will be critical. The technology is not suitable for tracking the behavior of individual resources.

Relevant Technologies – A number of commercial software products are available for the development of system dynamics simulations.

Possible Role of Standards – Standards for system dynamics modeling do not currently exist. While the basic notations (e.g., causal loops, stocks, and flows) are standard across system dynamics simulations products, there isn't standard for representing the model in a neutral format that can be input into different products. A technology is needed to allow large flexibility in modeling a wide range of systems. The downside of the flexibility is that models of the same phenomenon for the same system can vary widely when modeled by different analysts. Using standard modeling constructs to support classes of applications within a specific domain can help reduce the modeling effort and improve communication and understanding. Associated with such constructs, data standards will help further reduce the modeling effort, reduce modeling errors, and promote collection of data in forms suitable for supporting system dynamic models.

Expectations With Respect To Deliverables – For homeland security applications, new projects would be well advised to seek systems that are data driven and can be used at multiple sites. Alternatively, studies may be sought that utilize system dynamics simulation to analyze a policy issue and provide recommendations. In both cases, the deliverables should include documentation with model assumptions, major causal relationships, and verification and validation approaches used with their results. The study report should include scenario alternatives considered, relevant system outputs, and recommendations. For decision support systems, the software, associated user and system documentation, and application examples should also be included in deliverables.

Recommendations and Guidance – System dynamics simulation applications should generally be used for supporting policy decisions that have a large scope with impact over the long term. It is particularly applicable for studying the impact of alternate policies in social behavior, economy and finance, healthcare systems, and critical infrastructure domains (See [McLean 2009] for major domains for homeland security applications). Where possible the models should be made largely data driven to allow their application for multiple instances. Example applications include:

- A study of impact of pandemic influenza on the nation's population, infrastructure, and economy [NISAC 2007].
- Modeling the socio-political dynamics of Islamist movements [Mackerrow 2006].

For Further Information – For a general introduction to system modeling techniques, etc., see [Sterman 2000]. To learn more about systems dynamics modeling or to meet other professionals using this technology, investigate the annual conference [SDS 2011].

3.1.2 Discrete Event

Introduction – Discrete-event simulation (DES) is suitable for modeling system operations to evaluate system configurations and resource allocations in order to achieve desired system performance or to investigate causes of less than desired performance. It is generally used to model systems at medium to low levels of abstraction. DES models are generally used for planning purposes; however, they are increasingly used in near real-time decision support systems, particularly in manufacturing. Such applications are feasible in homeland security applications subject to availability of systems with real-time data.

In DES, the operation of a system is represented as a chronological series of events. As the name indicates, it uses discrete-event paradigm for representing time – the simulated clock time jumps from one event of interest to the next event of interest without going through successive unit increments. Beginning in the 1960s, discrete-event simulation gained popularity with availability of specialized languages.

Discrete-event simulation models may be developed using one of two major views: process view or event view. Process view essentially uses flow charts of process of interests and models them using corresponding simulation software features. The process view is also referred to as entity view or transaction view as it models the process that entities (or transactions) of interest go through in the system. The event view model is the actions that happen following an event. For example, consider modeling of treatment of victims of an emergency incident at the nearest hospital. The process view models the victim as an entity and the process the entity goes through - request an ambulance, be served by the emergency medical technicians (EMTs), be transported to the hospital, request emergency room resources, and be served by them. On the other hand, the event view models events that get triggered as needed. Arrival of a loaded ambulance at a hospital can be modeled as an event in an emergency response simulation. Upon arrival, emergency room staff (doctors, nurses, etc.) and resources (trauma treatment rooms) are assigned to treat the incident victim, while the ambulance and EMTs are released. The ambulance arrival event in the model is coded to model the actions of resource assignments and releases. The actions will also include setting up to trigger the treatment completion event after the modeled treatment time. At the treatment completion event, the modeled actions will include release of the doctors and nurses from serving the victim to becoming available for treating others.

DES models provide system performance measures such as system response and service times that are generally of interest to decision makers. Many DES commercial software allow development of detailed 3D animations that can help communicate the insights gained from simulation to people with limited background in the technology. However, unless thoughtfully managed, use of animations can easily mislead the audience.

Animation – is a dynamic graphic video of the events in the simulated system that can be used to show specific sequences to provide insights into the modeled operation. The animations can be 2D or 3D and at various level of details. Rich animations typically require large computing power and execution time.

Entities and Resources – are the items of interest in the system being studied. Entities generally go through a process and utilize or get served by resources while going through steps in the process. In the emergency room example, patients may be modeled as entities that go through the process while ambulances, EMTs, doctors, and nurses may be modeled as resources.

Event – is a change of a state in the modeled system. In the emergency room example, the arrival of ambulance at the hospital changes the state of the system – some resources (doctors, nurses) get busy while others become available (ambulance, EMTs) to serve other victims.

Queue – If the resources are not available, entities queue up waiting for service. In the emergency room example, non-critical victims may have to queue up waiting for doctors and nurses while they are tied up treating critical victims.

Clock – is used to represent time in the simulated world. An emergency response model may simulate two days of time following the occurrence of an incident. The simulated clock may be set to start at the time the incident happens and run for two days. As mentioned above the clock will jump from event to event. For example, if the ambulance leaves the incident site at 5pm and arrives at the hospital at 5:15pm and no other event of interest occurs during the time the ambulance travels, the simulated clock will jump from 5pm to 5:15pm. The simulation will model all the actions that happen at 5pm with the start of ambulance travel and then jump to 5:15pm and model all the actions that happen with the arrival of ambulance at the hospital.

Process – is the representation of the real-life sequence of related events that happen in the system of interest. In the emergency room example, the victim goes through the process of getting an ambulance, going to the hospital, and being served by the emergency room staff. This is captured in a process representation in the model.

Statistics – include the performance measures of interest that are useful in evaluating the system configuration, such as, resource utilization, service times, and their associated uncertainties. In the emergency room example, the utilization of hospital staff (doctors and nurses) and the associated average waiting times for the victims may be the key factor for decision-makers. If the doctors are very highly utilized, this might be reflected in long waiting times for the victims. The decision-makers can use the simulation model to trade-off the availability of staff and medical equipment resources (and corresponding costs) with acceptable waiting times for the victims.

Issues:

Potential Problem Area(s) – DES models generally include a lot more details than system dynamics models and hence require a lot more data and correspondingly large data collection efforts. Model development effort may also be larger than system dynamics models, though that may not be true if the domain has a good set of standards to support such efforts.

Relevant Technologies – A large number of commercial software products are available for discrete event simulation. Many of them are general purpose and can be used to model many types of operational systems, e.g., logistics, manufacturing, and hospital. Many are specialized to specific domains, such as for hospitals and medical processes. The prices of DES software vary widely depending on the capabilities and can range from $99 to $49000 [OR/MS 2011].

Possible Role of Standards – The recommended emphasis on seeking systems applicable across multiple sites points to data standards that will support such applications. Use of data standards for specific domain and applications (e.g., for configuring security check systems at different airports) can significantly speed up deployment of such applications and realization of associated benefits.

Expectations With Respect To Deliverables – For homeland security applications, projects should focus on acquiring implemented simulation models with animation capability and user interfaces for use across multiple sites. It is also possible to sponsor studies that utilize DES models for evaluating configuration of homeland security relevant operating systems such as airport security check systems. It is recommended that the emphasis be on the former rather than latter for good return on investment in this technology.

Integrating the acquired simulation models in a larger set of models through frameworks such as Standard Unified Modeling Mapping Integration Toolkit (SUMMIT) [SUMMIT 2011] should be considered, and if there is such potential, the application should provide interfaces to facilitate such integration.

Again, the deliverables should include documentation with model assumptions, conceptual representations of major processes modeled, and verification and validation approaches used with their results. The study report should include scenario alternatives considered, relevant system outputs, and recommendations. For decision support systems, the software, associated user and system documentation, and application examples should also be included in deliverables.

Recommendations and Guidance – DES model technology is quite mature and well-designed applications based on DES can help improve the understanding of system resource requirements, configurations, operations, and processes. The technology should be particularly used to evaluate plans across all the phases including prevention, preparedness, response, recovery, and mitigation. It is especially suitable for such applications in incident management, health care systems, and critical infrastructure domains of homeland security applications. Example applications include:

- Manpower planning and scheduling system for airports developed by Transportation Security Administration [Kalasky 2010].
- Modeling the response of a state health system to an infectious disease outbreak [Worth 2010].

For Further Information – For a general introduction to simulation including how discrete-event simulation works [Schriber 1998] and also for more advanced topics on Discrete Event Simulation see [Banks 2010]. For a discussion of Discrete-Events System Specification (DEVS), a formalism for modeling and analysis of discrete event systems, see [Zeigler 2000]. To learn more about DES and its applications, or meet other professionals and academicians using this technology, investigate the annual conference and its past proceedings [WSC 2011].

3.1.3 Agent Based

Introduction – Agent-based simulation (ABS) is suitable for modeling systems where the behavior is determined by the interactions of a large number of independent entities. Example applications include modeling the behavior of a crowd of people affected by an incident, and modeling the spread of a pandemic flu based on the behavior of individuals in the population in the affected area. ABS utilizes a decentralized representation of systems and allows the system behavior to be determined based on defined behaviors of a number of modeled agents. Agent-based simulations may follow the discrete-event paradigm or the continuous paradigm for time representation and they may utilize the hybrid form, i.e., using a combination of discrete and continuous representations.

The technology is based on defining individual agents with specified behavior. The agents may be used to represent people, organizations, equipment, etc. For homeland security applications, generally agents are used to model people. For example, to model the effect of an explosion on the crowd in the immediate area, people expected to move together may be represented as agents. Families that will move together and close-knit groups of friends can be represented as individual agents. Individual agents can also be used to represent single people who make decisions and take actions independently. Separate behaviors can be defined for different types of agents. An agent representing a family with young children may have the behavior of moving immediately in the opposite direction of the explosion albeit at a slower pace compared to an agent representing a family with teenaged children. An agent representing a young single person may have the behavior of moving towards the area of explosion to determine what happened and offer help to the victims. The overall crowd behavior will be determined based on aggregate effect of the behaviors of all the agents in the crowd during the simulation run. Associated with the above overview, the key terminology in ABS is briefly defined below based on [Macal 2011].

Terminology:

> **Agent** – is an entity with independent behavior based on its attributes and behavior rules. The agents have autonomy (independence), modularity (self-contained), sociality (interactions with other agents), and conditionality (state varying over time based on the situation). Agents may also be provided additional properties such as learning.

> **Agent Relationship**s – define how agents interact and who do they interact with. The modeled topology: none, a lattice/cell structure, 2D/3D Euclidean space, Geographic Information Systems (GIS), and networks may determine with whom they interact. For example in a lattice structure with squares, an agent may interact with agents in four cells that it shares the boundaries with, or eight cells including these and the four cells that it shares corners with.

> **Agents' Environment** – defines the modeling surroundings and the interactions of agents with the surroundings.

> **Emergence** – is the aggregate behavior that emerges based on the individual behaviors and actions of all the modeled agents. The overall system may develop self-organized sustainable patterns that haven't been explicitly programmed in to the system.

Issues:

> **Potential Problem Area(s)** – The primary problem with ABS models is the difficulty of validation. Social systems are quite unpredictable and their behaviors can change quite a bit depending on the individuals involved. The feasibility of the model outputs may be assessed. Predicting future outcomes of social systems based on ABS is fraught with risks.

> Similar to DES, ABS models may require a large amount of data for typical applications. In addition to the data, collection of knowledge to define agent behaviors and relationships is required.

Relevant Technologies – Both commercial and free research software tools are available for ABS. Professional licenses for commercial tools may cost around $15,000 per single user. ABS may also be developed using general-purpose DES tools defining agent behavior for specific events. ABS may also be developed using tools such as Excel and MATLAB or using general programming languages such as C++ and Java. General tools allow a lot of flexibility but require a large development effort while tools specialized for ABS provide many features that facilitate development but may constrain the flexibility.

Possible Role of Standards – Standard approaches for study of relevant scenarios using ABS would help facilitate their development and use for homeland security applications. This may include data interfaces and standard model constructs allowing use of different ABS tools to generate models based on the same data.

Expectations With Respect To Deliverables – Homeland security applications may focus on studies based on ABS to evaluate operational policies and procedures for systems involving social systems, such as large-scale evacuations, and response to pandemic flu. There may be limited opportunities for developing data driven ABS systems for application across multiple sites, though systems may be developed for providing insights to decision makers.

Recommendations and Guidance – ABS systems should be used where other simulation technologies are found to have limitations, such as, large-scale social systems that need to be modeled at a low abstraction level. It is suitable for applications involving modeling of social behavior aspects in domains such as incident management (e.g., crowd simulation), critical infrastructure (e.g., evacuations using transportation infrastructure), and health care systems (e.g., spread of infectious diseases). Examples of ABS applications include:

- Evaluation of counteractions against cyber-attacks [Kotenko 2007].
- Simulation of crowd behavior following an emergency incident [Shendarkar 2008].

For Further Information – For an introduction to agent-based simulation and its applications see [Axelrod 1997]. A good introductory tutorial is provided by [Macal 2011]. For a view of leading applications and research, visit the annual Winter Simulation Conference [WSC 2011] or the annual Agent Directed Simulation Symposium [ADS 2011].

3.1.4   Physical-Science-Based

Introduction – Physical-science-based simulations utilize scientific knowledge, e.g., the laws of physics or mathematical models of observed phenomena to study, understand, or predict the behavior of physical systems. Physical systems can range from very simple, e.g., the study of motion of a bullet, to very complex, e.g., the behavior of organisms, crowds, or global climate.

Physical-science-based models may use mathematical equations and schematic diagrams as conceptual models. Physical-science-based models typically utilize differential equations based on laws of physics that model such factors as mechanical dynamics and statics, material behavior under stress and impact, and fluid dynamics. They are generally used for modeling at detailed level, that is, at low abstraction level, such as, equipment and equipment component behavior, behavior of built structures when subjected to explosions in close proximity, and spread of fire through buildings. There are also a few instances of their use at high abstraction levels such as for modeling weather systems. Physical-science-based simulations use the continuous paradigm for representing time.

The category of physical-science-based simulation covers a wide variety of systems and requires domain specific expertise. For example, it includes simulation of atmospheric releases requiring expertise in computational fluid dynamics, meteorology, etc., and it includes behavior of electronic devices subjected to rough environments requiring expertise in semiconductor physics, signal noise, frequency conversion, etc.

Physical-science-based models in materials science may be classified based on the scale of phenomenon into atomic, micro, meso, and macro [Steinhauser 2008]. The model may use quantum mechanics and microstructure simulations at the nanoscale and coarse-grained atomistic simulations, and use classical particle approach to solve Newton's equations of motion at the meso- and macro-scale.

Terminology:

Rigid body – is a representation of a solid body of finite size in which deformation is assumed to be negligible regardless of the external forces applied on it. This provides an approximation for physical-science-based simulation when modeling kinematics and dynamics of bodies including collisions, contacts, and friction and such phenomenon as shock propagation.

Particle systems – is an approach that involves modeling of movement of individual particles when subjected to physical changes and forces. This approach is commonly used for modeling volumetric effects such as fires, explosions, smoke, or fluids. Particle-particle interactions are modeled for more complex systems.

Elasticity – is the tendency of a body to return to its original shape after it has been stretched or compressed. This property may be modeled when the rigid body representation is not appropriate.

Deformable and plastic bodies – are bodies that experience a permanent change in their structure due to physical changes including forces.

Fluids, gases, and complex materials – require smoothed particle hydrodynamics for simulation of viscous fluids and complex materials.

Issues:

Potential Problem Area(s) – Development of models requires high domain expertise with deep understanding of the involved physics. The ability to validate physical-science-based models depends on the scale of phenomenon modeled. Simulations of large structures may require expensive experiments for validation. Scaled down models of reality may have different physics and thus pose additional challenges for validation.

Relevant Technologies – A wide range of commercial and research software exists for physical-science-based simulations. For example, a number of Finite Element Modeling (FEM) software are available that can be used to analyze behavior of material components under stress. Models for physical-science-based simulation of specific equipment may need to be custom-developed using tools for that domain.

Possible Role of Standards – Standards will help facilitate the development of models within specified domains. For example, standard ways to specify 3D design of aircraft and ships will reduce the effort to model their behavior.

Expectations With Respect To Deliverables – Projects involving physical-science-based simulation may deliver studies or reports evaluating specific structures in most cases and software in some cases. For example, impact of explosions on aircraft or ships will require building customized detailed models of specific aircraft and ships for the study.

Recommendations and Guidance – Physical-science-based simulations should be used for phenomenon where physics of the process is critical such as for examining blast resistance of structures, and predicting the dispersion of atmospheric releases (plumes). Some examples of suitable applications for modeling include:

- Hazardous material releases (e.g., predicting the dispersion of plumes).
- Critical infrastructure (e.g., impact of explosions on structure).
- Incident management (e.g., devastation caused by explosions in a crowded area).
- Health care (e.g., virus interactions with healthy cells in the human body) and other systems, equipment, and tools (e.g., performance of radiation scanners).
- Wildfires [Younker 2002].
- Explosions inside containers used to dispose of improvised explosive devices (IEDs) [LANL 2005].

For Further Information – Sources for further information are highly dependent on the physical phenomenon of interest. For more on physical-science-based simulations, see [Steinhauser 2008], [Thalman 2008], [Anderson 2008], [Engquist 2009], [Gen 2008], and [Bonate 2006].

3.1.5   Gaming Systems

Introduction – Gaming technologies (e.g., video game technologies) may be very useful in meeting training and exercise needs in the incident management and healthcare systems domains, as well as other areas. Game technology promises to provide a more engaging simulation-based learning experience than traditional classroom or computer-aided instruction methods. Video game engines provide integrated environments for creating virtual worlds. They typically have capabilities for creating three-dimensional graphics, sound, animated characters, intelligent character behaviors, and various physical phenomena. They may also provide support for the creation of user level game modifications, web-based software distribution, and distributed game participation over the Internet.

Game technology has primarily been used for entertainment rather than educational purposes in the past. A number of changes to gaming systems will be required to support DHS needs. New functionality needs to be incorporated into game software to make it suitable for training applications, e.g., security mechanisms and interfaces to learning systems software (discussed later in this document).

Successful development of games for homeland security will require considerable forethought as well as expenditure of resources. Software licenses for game development systems and mass distribution of game software are often quite expensive. Pervasive use of this technology will require that many contractors access licenses to develop simulation-based learning applications. Perhaps hundreds or thousands of game-based training applications will ultimately be distributed. Game engine developers often collect royalties on each game sold. If commercial game engine software is used for DHS training and exercise applications, the traditional business models of game engine vendors may need to be adapted since DHS is not expected to sell games or be in the business of distributing games for profit.

Game software is typically very complex. Some of the key elements, i.e., modules that are typically found in gaming systems include:

- Graphics and sound.

- Game engine management.
- Script compilers/interpreters.
- Character animation.
- Physics engine.
- Artificial intelligence.
- Game editor and development tools.
- Application data import and export.
- Multi-player operations and server support.
- Software distribution and security mechanisms.

Game-based simulations are typically complex software systems that are subject to a number of technical and management risks. Considerable up-front thought and analysis should be given to the issues associated with game development. Some of the questions that should be asked:

- Is a game the best way to achieve this objective?
- What type of game best satisfies needs?
- Will the game best be implemented as a two-dimensional board game or a three dimensional virtual reality environment?
- How will the game be used and where will it be installed?
- How many users (players) are expected?
- Will there be non-player characters?
- Will artificial intelligence be required to implement player or non-player characters?
- How long will game sessions typically last?
- How much graphical realism is appropriate?
- Will there be sound?
- What game assets will be needed?
- Will physics be needed to model the behavior of objects?
- Are cut sequences planned (i.e., game action stops while videos play to pass relevant information to players)?
- Will the game need to interface to other homeland security software applications?
- What types of modifications are expected will be allowed?
- Will sensitive data be involved, e.g., local responder resource information?
- What types of security mechanisms are needed to protect sensitive or proprietary data?
- What sorts of game engine or software libraries will be used?
- Do the developers have the appropriate technical expertise and experience to develop the game?
- What are the risks associated with staff turnover and how will they be handled?
- Will the developers provide mockups of the game for approval before a development go-ahead is given?
- How long will it take to develop the game?
- What are expected development costs?

A final thought – Negative training must be avoided. "Negative training is a process in which knowledge, skills and/or attitude are changed in such a way that when presented with the real task the student, if following what they have been taught, would carry out the process incorrectly or, in the worst case, dangerously." [Lever 2004]

Asset – is any of a number of objects used in the creation of video games, including such things characters, tools, vehicles, buildings, sounds, and animations.

Role-playing game (RPG) – is a game genre in which each participant assumes the role of a character. The game is generally in a computer generated setting, which can interact within the game's imaginary world to achieve some objective(s) or quest(s). Players issue commands to one or more characters whose abilities typically increase with experience and achievements as the game is played. RPGs tend to offer complex and dynamic character interaction based upon artificial intelligence and scripted behavior of computer-controlled non-player characters.

Game modifications or mods – "Mod or modification is a term generally applied to personal computer games (PC games), especially first-person shooters, role-playing games and real-time strategy games. Mods are made by the general public or a developer, and can be entirely new games in themselves, but mods are not standalone software and require the user to have the original release in order to run. They can include new items, weapons, characters, enemies, models, textures, levels, story lines, music, and game modes. They also usually take place in unique locations." [Wiki 2011b]

Issues:

Potential Problem Area(s) – Game development is an exciting area that many software developers may be anxious to jump into without investing appropriate time and resources in up front analysis, modeling, and documentation. Sponsors/managers must be extremely vigilant to ensure that appropriate analysis is completed before software development begins.

"Not invented here" is a familiar phrase in the software-development community. It refers to a reticence on the part of developers to use software that has been created by others, and their desire to re-create code that already exists (thus resulting in redundant efforts and waste). Considerable thought should be given by sponsors/managers to address this problem, e.g., carefully apportioning work to different developer organizations. For example, one developer might develop game asset libraries to be used by all. Another might create the execution environment. A third might create user level game software, e.g., scenarios, settings, and scripts.

Another issue that was briefly introduced above is the cost associated with distributing games among the user community. Deployment expectations and licensing costs should be considered before any commitments to game development and execution environments are made.

Relevant Technologies – There are a number of commercial and open-source game development environments, execution engines, and software libraries currently available. Any new game development efforts should not be undertaken until the use of available tools has been sufficiently explored.

Possible Role of Standards – A number of standards exist that may be applicable to the development of games for homeland security training and exercises. The standards include those for scripting languages, geographic information system (GIS) data, de facto commercial standards for game assets (e.g., 3D Studio Max graphics models), graphical programming languages, and software subroutine libraries.

Expectations With Respect To Deliverables – It is likely that for almost any game that is created, there will be a need to change scenarios, characters, scripts, locations, and resource data, e.g., responder capabilities. In game terminology, the ability to extend or adapt a game is typically provided in a game modification or mod capability. Deliverables for any game should include a mod capability, as well as example test data, and appropriate documentation for both users and developers.

Recommendations and Guidance – A long-term strategy for the development and usage of game-based homeland security training and exercises would help guide future efforts in this area. Without such a strategy, considerable resources may be unnecessarily expended and opportunities may be lost due to duplicative or incompatible development efforts. The strategy should address issues such as:

- Types of training/exercise applications that are needed/expected.
- Game development processes.
- Use of proprietary versus open architectures and source code.
- Intellectual property rights.
- Establishment of shared libraries for game assets, scenario data, and software.
- Use of standards.
- Incorporation of security mechanisms.
- Validation and testing of applications.
- Distribution/deployment plans.
- Documentation policies.

It may be appropriate to collaborate with other interested government agencies to develop a common strategy for gaming technology.

For Further Information – For information on management and software engineering practices for game development (including conceptual modeling, detailed design, staffing, and development processes), see [Bethke 2003] and [Flynt 2005]. For discussions of the game development philosophy, thought processes, and cultural issues, see [Salen 2004].

3.1.6 Tabletop Exercise Enablers

Introduction – The FEMA Homeland Security Exercise and Evaluation Program (HSEEP) is a capabilities and performance-based exercise program, which provides a standardized policy, methodology, and terminology for exercise design, development, conduct, evaluation, and improvement planning. HSEEP defines two types of exercises, i.e., discussion-based and operations-based. The HSEEP provides a brief summary of exercise types as follows:

Discussion-based Exercises familiarize participants with current plans, policies, agreements, and procedures, or may be used to develop new plans, policies, agreements, and procedures. Types of Discussion-based Exercises include:

- Seminar. A seminar is an informal discussion, designed to orient participants to new or updated plans, policies, or procedures (e.g., a seminar to review a new Evacuation Standard Operating Procedure).
- Workshop. A workshop resembles a seminar but is employed to build specific products, such as a draft plan or policy (e.g., a Training and Exercise Plan Workshop is used to develop a Multi-Year Training and Exercise Plan).

- Tabletop Exercise (TTX). A tabletop exercise involves key personnel discussing simulated scenarios in an informal setting. TTXs can be used to assess plans, policies, and procedures.
- Games. A game is a simulation of operations that often involves two or more teams, usually in a competitive environment, using rules, data, and procedure designed to depict an actual or assumed real-life situation.

Operations-based Exercises validate plans, policies, agreements and procedures; clarify roles and responsibilities; and identify resource gaps in an operational environment. Types of Operations-based Exercises include:

- Drill. A drill is a coordinated, supervised activity usually employed to test a single specific operation or function within a single entity (e.g., a fire department conducts a decontamination drill).
- Functional Exercise (FE). A functional exercise examines and/or validates the coordination, command, and control between various multi-agency coordination centers (e.g., emergency operation center, joint field office). A functional exercise does not involve any "boots on the ground" (i.e., first responders or emergency officials responding to an incident in real time).
- Full-Scale Exercises (FSE). A full-scale exercise is a multi-agency, multi-jurisdictional, multi-discipline exercise involving functional (e.g., joint field office, emergency operation centers) and "boots on the ground" response (e.g., firefighters decontaminating mock victims). [HSEEP 2011]

A number of documents are defined that are associated with the implementation of most exercises: Situation Manual (SitMan), Exercise Plan (ExPlan), Controller and Evaluator (C/E) Handbook, Master Scenario Events List (MSEL), Player Handout, Exercise Evaluation Guides (EEGs), and After Action Report/Improvement Plan (AAR/IP).

The MSEL is a chronological timeline of expected actions and scripted events (i.e., injects) to be inserted into operations-based exercise play by controllers to generate or prompt player activity. It ensures necessary events happen so that all exercise objectives are met. For large exercises, the management of a large number of events in the MSEL and the capture of data about the exercise can be overwhelming. Functions that may be performed by automated TTX software include:

- Provide web-based interface to inject management system to enable remote access to and operation of the exercise control system over the Internet.
- Provide menus and forms for creating injects, adding/deleting from library, selecting for a particular exercise, and setting inject parameters (e.g., source, target addressee, video or audio stream/channel, and inject variables).
- Provide controls for identifying exercise and key parameters, starting, pausing, initiating rollbacks, time warp (jumping clock ahead), and stopping inject sequences.
- Provide exercise control functions and time clock functions for table-top management exercises.
- Maintain database of exercise scenarios, alerts, messages, and decisions.
- Dispatch messages to exercise control personnel.
- Record message consequences note by controllers.
- Provide editing functions for development of exercise scenarios.
- Permit user-friendly scenario development (i.e., via Excel Spreadsheet template or import modules)
- Deliver Common Alerting Protocol (CAP) compatibility.
- Automation of phone calls and audio playback, transmission of video streams, and passing of messages and scripting instructions to other exercise systems, Emergency Data Exchange Language (EDXL) compatibility, seamless integration with the Common Operating Picture.

NIST implemented an Exercise Control System (ECS), which performs many of these functions. The system was used in the California Golden Guardian Exercise of 2007 [NIST 2007].

Terminology:

Master Scenario Events List (MSEL) – "A MSEL lists all exercise messages and key events in a table that specifies the time the message is expected to be delivered, who delivers it to whom, a message number, and a short description of the message. Some MSELs also contain the anticipated responder actions and associated exercise objectives to assist the controllers and evaluators in performing their functions. The MSEL identifies the timing and summary content of all key events, messages or injects, contingency messages, and anticipated responder actions for the duration of the exercise." [ORAU 2011]

Inject – is information in the form of a text message, phone call, radio transmission, or audio/video stream that is entered into an exercise at a specific point in time, typically to provoke actions/responses on the part of exercise participants.

Issues:

Potential Problem Area(s) – Large, multi-organization TTXs may have very large and complex MSELs. TTX support tools need to provide capabilities to help make the development of new exercises based on old ones more efficient. Tools need to provide functionality for the management, parameterization, and re-use of chunks of interrelated injects to minimize the amount of manual editing of MSELs that is required. Injects may typically be complex structures. Although there are some standards that can be used to format injects, more will probably be needed.

Relevant Technologies – A number of commercial and prototype tools are available to support the development of tabletop exercises, e.g., the Oak Ridge Exercise Builder, the NIST ECS, as well as others.

Possible Role of Standards – A number of standard message formats have been developed that can be used for TTX injects, e.g., EDXL and the Common Alerting Protocol (CAP). Standard web technologies can be used to automate table-top exercises at multiple locations using the Internet.

Expectations With Respect To Deliverables – TTX enablers should be well documented to support their widespread use. Standards that have been used should be clearly identified. Test data sets based on real exercises should be provided as examples to simplify the creation of new exercise scenarios.

Recommendations and Guidance – Development of a public domain, i.e., freely distributable tool for conducting table-top exercises using Web technology, similar to the NIST ECS prototype, could reap widespread benefits given the number of TTXs that are conducted across the nation each year. Such a tool would need to be easy to use and would need to incorporate additional security features to protect sensitive exercise data.

For Further Information – For more information on the Homeland Security Exercise and Evaluation Program (HSEEP) including policies and procedures, design of exercises, and document formats, see [HSEEP 2011]. For examples of past tabletop exercises, see [CSIS 2011].

3.2   Development Technologies

This section introduces two development technologies relevant to the creation of homeland security M&S applications, namely:

- Simulation environments and languages.
- Learning Content Management Systems (LCMS) and Learning Management Systems (LMS).

Each of these development technologies is presented in detail below.

3.2.1   Simulation Environments and Languages

Introduction – A number of simulation environments and languages across the multiple categories of simulation technologies are presented in section 3.1. Most of the simulation environments and languages available today started as specialized procedural programming languages that required modelers to write lines of codes. Over time the majority of simulation languages developed for creating and supporting simulation environments with graphic user interfaces (GUIs). GUIs allow developing models by assembling icons representing different modeling constructs and populating their parameters to represent the real-world system or design being modeled. Many simulation software packages do allow advanced modelers to code in the language underlying the graphic modeling constructs. For many of the simulation environments, the underlying language is an evolved form of the original simulation language developed decades ago. In addition to the facility of programming in the underlying proprietary language, many software environments allow interfacing routines written in languages such as Visual Basic for Applications (VBA), C++, and Java. The discussion below follows the order in which simulation technologies were presented in section 3.1.

Jay Forrester developed a system dynamics simulation approach at MIT in the 1950s. A group of researchers working with him developed DYNAMO (DYNAmic Models), a language for system dynamics simulation. Current leading system dynamics environments are generally descendants of DYNAMO but all have adopted graphic user interfaces allowing users to construct causal loop and stock-and-flow diagrams and populate associated parameter windows with details of the relationships. These environments include STELLA, i-Think, PowerSim and Vensim.

Initial development of discrete-event-simulation languages can also be tracked to the late 1950s and early 1960s with the two earliest ones being General Simulation Program (GSP) developed in UK and General Purpose Simulation System (GPSS) developed in the US. Three main streams of discrete-event-simulation languages developed in the early years based on the world views: activity scan, process view, and event view. Over time the latter two views became prominent, with process view languages including GPSS, its descendants, and SIMULA, and event view languages including SIMAN, SLAM, and SIMSCRIPT. Again, over the years most discrete-event-simulation languages transformed into simulation environments with graphical user interfaces allowing users to assemble icons representing different modeling constructs. The users generally have the flexibility to drop into the underlying language for modeling unique situations that are not covered by generic modeling constructs. Current general purpose discrete event simulation software (with their underlying languages identified in parentheses) include ARENA (SIMAN), AWESIM (SLAM), GPSS-PC (GPSS), and SIMSCRIPT III (SIMSCRIPT). Other general-purpose simulation software with their own proprietary languages include AutoMod, ProModel, Enterprise Dynamics, and SIMUL8. The more recent commercial simulation software offerings have been all simulation environments, with the only exception being SLX (Simulation Language with eXtensibility). SLX is a simulation language developed in recent years to provide a large flexibility to modelers and to provide capabilities that were lacking in earlier simulation languages.

Recent DES environments and languages have been moving closer to implementing object-oriented structures.

Agent-based-simulation software appeared much later compared to system dynamics and discrete-event-simulation languages. The first such software, Swarm, was launched in 1994 by Santa Fe Institute. ABS software for large-scale developments also includes Repast and MASON. Environments that provide ABS capabilities in a desktop include Repast Symphony – a version of Repast that uses point-and-click interface and NetLogo. (For information on NetLogo, see [CCL 2012].) Object oriented languages lend themselves well to ABS with agents and their behavior being defined as object classes.

There are a few simulation software packages that provide capabilities for building SD, DES, and ABS models; or models that combine the three technologies. Such software includes AnyLogic, ExtendSim, GoldSim, and Simio. Among these Simio is the newest environment and is completely object-oriented.

A wide variety of physical-science-based-simulation tools exist catering to different domains related to homeland security applications. Rigid body dynamics tools include Adams, DynaWiz, Umbra, and Visual Nastran 4D. Finite element method analysis and numerical simulation environments for elastic-material/soft-body physics analysis include Abaqus, ANSYS, DYNA3D, FEMDesigner, FEM3MP and LS-DYNA. Computational Fluid Dynamics (CFD) environments include FLOW-3D, ANSYS-CFX, FLUENT, STAR-CD, and FEMLAB. Many of the physical-science-based simulation tools are highly specialized languages and exist in the form of proprietary codes used at national laboratories.

Terminology:

> **Object Oriented Programming (OOP) language** – OOP languages allow defining classes and objects corresponding to real-world object or concepts together with the parameters and methods associated with them. A key feature of OOP languages is inheritance – allowing objects within the same class to inherit parameters and methods from parent objects. Most of the recent general purpose languages such as C++ and Java, simulation environments such as Simio, and special purpose languages such as SLX are object oriented.

> **Procedural programming language** – Procedural programming languages require coding the logic in a step-by-step sequence of statements that follow the syntax of the particular language. Programming languages such as Fortran, Cobol, Basic, and C are procedural languages.

Issues:

> **Potential Problem Area(s)** – Simulation software and languages have subtle differences that need to be understood by the modelers and analysts to allow them to model correctly. Some of the representation errors may be caught during Verification and Validation of the models but others may slip by affecting the simulation results and in some cases, the decisions made based on the results. An example of a subtle difference among DES software is the way they resolve time ties, i.e., the order in which the events scheduled for the exact same time are executed (see [Schriber 2011] for more details.)

> **Relevant Technologies** – NA

> **Possible Role of Standards** – Standard ways to define models for each technology that help improve communication among different models and in turn, their validation. For example, the Discrete-Events-System Specification (DEVS) is a modular and hierarchical formalism for

modeling and analysis of discrete-event systems. DEVS supports all discrete-event models regardless of the simulation language or environment used.

Expectations With Respect To Deliverables – NA

Recommendations and Guidance – Selection of the simulation environment or language should generally be left to the modelers or analysts since they develop expertise in one or a few modeling environments or languages over the years. It is important that the modelers understand the subtleties of the simulation environment or language to correctly model the subject phenomenon and hence the need to have the expertise in the environment or language. The above recommendation is for the environment or language within each technology; the system dynamics simulation environment or language shouldn't be used for decisions that are appropriately addressed using DES technology just because the modeler has the associated expertise.

For Further Information – See software survey articles and introductory tutorials for each simulation technology: [Wiki 2011c] for system-dynamics simulations, [OR/MS 2011] and [Schriber 2011] for DES, [Macal 2011] for ABS, and [Trinkle 2011] for rigid body dynamics simulation software, a subset of physical-science-based simulation software.

3.2.2 Learning Content Management Systems (LCMSs) and Learning Management Systems (LMSs)

Introduction – Simulation could be a very useful tool for training in the incident management, healthcare, as well as other domains. Simulations may be decomposed into reusable learning objects (RLOs) that can be embedded in courses for classroom or individual training. If simulations are used for individual or classroom training exercises, LCMS and LMS technology is appropriate for the storage and management of simulation modules and data. The LCMS provides a central repository for the storage and retrieval of learning content objects. The LCMS may include Learning Management System (LMS) and Course Authoring System (CAS) functions. Major functions of the LCMS may include:

- Serve as a centralized repository for learning objects – simulations may be learning objects.
- Provide LMS functionality.
- Provide CAS functionality.
- Provide network connectivity and interact with external LMSs to deliver the course content to learners.
- Provide network connectivity or media input options for external CAS objects.
- Provide a user interface for administering the repository including learning object tagging, cataloging and indexing, and searching.
- Manage meta-data that describes the relevance of each learning object for specific learning objectives.
- Provide authoring functions for creating learning objects in standard learning object formats.
- Import and convert learning objects authored external to the system into learning object formats.
- Provide authoring capabilities to create courses from the learning objects contained in the repository.
- Manage workflow for learning object content creation.
- Dynamically assemble personalized courses based to meet specific learning objectives.

Major functions of the LMS include:

- Publish a catalog of course offerings.

- Provide communications connectivity and interact with the LCMS to obtain learning object course content.
- Provide access control to courses including mechanisms for course enrollment and checking of prerequisites.
- Manage personalized learning plans.
- Launch and track learning applications.
- Provide instructor interface for grading, retaking courses, and setting up courses.
- Maintain transcripts for the student population.
- Take required tests and assessments online.
- Track student progress, scores, completion, etc.
- Enable collaboration and communication between students and instructors.
- Enable virtual classrooms.
- Provide an application programmer interface (API) for interacting with courseware.
- Launch and interact with executable course content.
- Identify educational plans, measure and map skills, and perform gap analysis.

Terminology:

Learning Content Management System (LCMS) – The LCMS is a development environment where multiple learning content authors can create, store, reuse, manage, and deliver digital learning content.

Learning Management System (LMS) – The LMS is a system for managing learners (i.e., users of simulations in classroom or individual training environment) and keeping track of their progress and performance across all types of training activities.

Reusable Learning Object – "Object or set of resources that can be used for facilitating intended learning outcomes, and can be extracted and reused in other learning environments. Associated with e-learning resources that can be used in multiple learning environments." [E-Learning 2011]

Issues:

Potential Problem Area(s) – M&S applications are currently developed as stand-alone, monolithic systems. Systems developed as monolithic systems may not be suitable for classroom or remote training of homeland security personnel. Using portions of M&S applications to support classroom or remote training will most likely require restructuring of software. DHS should work with instructors and personnel familiar with learning environments to establish requirements, where appropriate, for the development of simulation-based learning objects. Also, there is no consensus agreement of a precise definition for the learning object concept.

Relevant Technologies – There are a number of commercially available LCMS, LMS, and CAS products available in the market today.

Possible Role of Standards: The Advanced Distributed Learning Initiative (ADL) is responsible for the standardization of the Shareable Content Object Reference Model (SCORM). The SCORM defines a Web-based learning "Content Aggregation Model" and "Run-Time Environment" for learning objects. SCORM is a collection of specifications adapted from multiple sources to provide a comprehensive suite of e-learning capabilities that enable interoperability, accessibility and reusability of Web-based learning content. The SCORM includes aspects that affect learning management systems and content authoring tool vendors, instructional designers and content developers, training providers and others. For further information on ADL and SCORM, see [ADLNET 2011].

Expectations With Respect To Deliverables – Simulation objects used in courses at homeland security training facilities should be delivered in a SCORM-compliant format so that they can be directly imported into LCMS and LMS systems.

Recommendations and Guidance – Needs and requirements for LCMS, LMS, and CAS systems should be established. Studies should be conducted to determine how to best modularize simulation objects for use/reuse in learning environments. Policies for the acquisition of LCMS and LMS software should be established by DHS. Systems meeting needs and requirements should be identified through a review of commercially available products. Such a review might be conducted with other agencies that are in the process of purchasing similar systems. Preferred systems should be identified to minimize the proliferation of systems through training facilities and possibly obtain cost savings through multiple purchases.

For Further Information – Information on SCORM-compliant LCMS, LMS, and CAS products can be found at [ADLNET 2011]. A comprehensive set of definition of e-learning terminology can be found at [E-Learning 2011].

4 Simulation Application Components

There are several key elements that should comprise most homeland security M&S applications. They are respectively:

- Simulation engine.
- User interfaces.
- Input and output data types.
- Databases, data files, and translators.
- IT Security mechanisms.

This section provides a brief overview of these elements.

4.1 Simulation Engine

Introduction – The construction of a simulation usually involves some sort of software development. Since software development is often a costly, time-consuming, and error-prone process, minimization of programming and re-use of validated code is highly desirable. A certain amount of code re-use can be achieved through the utilization of *simulation engines*. These *engines* or *simulators* are computer programs that typically provide functions to:

- Develop and manage simulation models.
- Execute the simulation.
- Assist in model debugging.
- Incorporate programming language extensions.
- Input and output data.
- Initiate and terminate simulation runs.
- Generate statistical variations between runs.
- Create and display 2D or 3D visualizations.
- Analyze results.
- Produce reports.

The functionality for execution of the simulation, of course, depends on the simulation technologies that the simulation engine provides. For discrete-event simulation, the simulation engine should provide some of the following capabilities.

- Initializing and maintaining the simulation clock.
- Ability to define statistical distributions to represent stochastic phenomena.
- Using random number generators for sampling for statistical distributions.
- Maintaining list of current events and executing the events that are ready.
- Maintaining list of future events and transferring ready events to current events list as the simulated clock advances.
- Generating and moving the entities through the model.
- Maintain attributes of the entities.
- Destroy the entities.
- Maintain variables that model resources.
- Manage and track resources states such as busy, starved, blocked, and set up, as appropriate for the resource type.
- Maintain queues of entities waiting for the needed resources to become available.

- Provide the ability to implement different priority disciplines for the queue.
- Maintain calendars for the availability of resources.
- Collect statistics such as wait times, service times, and resource utilization.
- Reporting aggregate statistics.
- Generate event traces that can be used for concurrent or post simulation animation.

By using a simulation engine, much low-level coding could be avoided. Unfortunately, most simulation engines currently run as stand-alone systems on PC type computers. As such, new systems or modifications to existing systems will be required before simulation engines will be able to run as part of large distributed simulation environments, browser plug-ins, mobile code, or run remotely on a server and interact with browser plug-ins.

There are a few simulation engines available, as opposed to simulation environments discussed in section 3.2.1, that can be embedded in applications such as a decision support system. [Schwetman 2001] describes CSM19, a discrete-event-simulation engine that is available commercially for embedding into applications. Programmers can write C++ programs that use CSM19 library for simulation functionality. [Lanner 2011] has developed L-SIM DES simulation engine for embedding into third-party applications for business-process modeling. Similarly [Powersim 2011] provides an SD simulation engine for building custom applications. PRIME, an open source distributed simulation engine, allows setting up and running simulations using available parallel and distributed processors [Liu 2009]. PRIME has been used by the National Infrastructure Simulation and Analysis Center (NISAC) as the platform for SimCore, a discrete event simulation framework, which in turn has been used to build agent-based simulations [NISAC 2011].

Terminology:

> **Physics Engine** – is a software system that provides approximate simulation of physical systems such as rigid body dynamics, elastic/soft body dynamics, particle systems, and fluid dynamics. A number of physics engines exist for supporting gaming that may use approximations in the interest of real-time performance. Engines for serious applications are sometimes referred to as high-precision physics engines and such engines should be suitable for homeland security applications.

> **Game Engine** – is a software system designed for developing video games that provides many of the capabilities that are required for the purpose. These engines generally provide an integrated development environment facilitating the creation of games and provide capabilities such as graphics, sound, physics, and artificial intelligence functions.

Issues:

> **Potential Problem Area(s)** – A number of simulation engines are available for use in gaming applications. Many of these simulation engines are focused on creating a feasible scenario that may or may not be technically correct. Modelers selecting simulation engines for embedding in serious applications hence need to ensure that the engine is intended for applications requiring technically correct simulations. Also, decision-support systems based on utilizing commercial simulation engines may incur large costs for multi-station or multi-user licenses if needed for deployment.

> **Relevant Technologies** – A wide range of hardware configurations exist for executing simulation engines. These configurations include executing standalone on desktop or workstation, client-

server set-ups, and executing via the web (using web-services or applets). Simulation engines may be capable of exploiting parallel and distributed architectures such as shared memory multi-processor machines or grids. Challenges for running models developed using commercial-off-the-shelf (COTS) simulation engines and software are discussed in [Taylor 2011].

Possible Role of Standards – Defining standards for simulation engines that exploit multiple processor machines may encourage competition and in the long term reduce the licensing cost for large deployments.

Expectations With Respect To Deliverables – Homeland security applications should be developed based on simulation engines that do not pose large costs and challenges due to licensing issues. Contracting arrangements should be set up such that deployment at multiple locations or use by multiple users minimizes licensing expenses. Also, future applications should look for simulation engines capable of employing multiple processors for execution speed advantages.

Recommendations and Guidance – It will serve DHS well to look for applications built on simulation engines that provide technically correct answers, are capable of exploiting multiple processors, and do not have expensive licensing arrangements for large deployments. It should be noted that for some training applications, the need for real-time experience may lead to accepting simulation engines that provide feasible scenarios that may not necessarily be technically correct.

For Further Information – For an overview of current state of the art for executing simulations using multiple processors see [Taylor 2011]. [SISO 2010] describes a standard for COTS simulation package interoperability that will help in integrating multiple simulations to put together large models.

4.2   User Interfaces

Introduction – Another key component of an M&S application is the human user interface it provides. User interfaces are the various visual displays, audio inputs and outputs, and in some cases mechanical devices, e.g., computerized medical mannequins that associated with an M&S healthcare application. Depending on the type, size, and intended distribution of an M&S application, a number of different types of user interfaces may be required. User interfaces determine who can interact with the application and the types of roles they may play, as well as how easy or difficult it may be to learn and use the application.

Some possible categories of interfaces for M&S applications include *Manager/Analyst, Systems Engineering and Support Staff, Instructor and Trainer, System Administrator, Exercise Management, First Responder, Incident Management, Support Institution Staff, Civilian Population,* and *Opposing Forces*. Many specific interfaces are needed to support M&S applications used for training or exercises.

Manager/Analyst may be used to setup simulation study parameters, run simulations, access results, and generate reports. *Systems Engineering and Support Staff* is used to develop, modify, and control M&S applications and for executing systems engineering processes. *Instructor and Trainer* is for configuring simulation-based instruction, initiating exercises, controlling flow, interrupting the execution, modifying execution parameters, and tracking student progress. *System Administrator* is used to deal with issues such as installation versions and linking with databases and other systems. *Exercise Management* is used to control the simultaneous execution of multiple M&S applications and interactions with human participants in the exercise. *First Responder* allows the performance of front-line roles at an incident site including fire fighting, crowd control, victim triage, and terrorist capture. *Incident Management* is used for training of decision-makers in incident command center and emergency operations centers at various levels of hierarchy. *Support Institution Staff* is for organizations, such as hospitals, that provide disaster support. *Civilian Population* allows role-playing for civilians caught up in an emergency incident. *Opposing Forces* is used for playing the role of terrorists or enemy combatants unleashing an attack.

Ben Schneiderman, an expert on user interface design, has suggested the following criteria for evaluating the usability of a user interface:

- Learnability: How easy is it for users to accomplish basic tasks the first time they encounter the design?
- Efficiency: Once users have learned the design, how quickly can they perform tasks?
- Memorability: When users return to the design after a period of not using it, how easily can they reestablish proficiency?
- Errors: How many errors do users make, how severe are these errors, and how easily can they recover from the errors?
- Satisfaction: How pleasant is it to use the design? [Wiki 2011d]

Commercial M&S applications will typically include predefined tools and functions for constructing user interfaces. As such, analysts may be limited in flexibility as to the interfaces that they can create.

Terminology:

Graphical user interface (GUI) – a visual mechanism for interacting with a computer or computer program using components such as desktop displays, windows, icons, menus, pointers, and pointing devices.

Issues:

Potential Problem Area(s) – In the absence of DHS user interface guidance or standards, it is likely that each M&S application developed by a different project team will have a unique user interface. The diversity of user interfaces will lengthen the learning curve for M&S users as they move from application to application. M&S applications based on commercial environments may limit the ability of an analyst to customize a user interface.

Relevant Technologies – Many commercial and public domain software tools and libraries are available to create graphical user interfaces for software applications.

Possible Role of Standards – A number of standards exist that relate to the development of user interfaces, e.g., ISO TC 159/SC 4 has a number of standards [ISO 2011a] on human-system interaction, and ISO/TR 16982:2002 Ergonomics of human-system interaction – Usability methods supporting human-centered design [ISO 2011b].

The Federal Government has mandated that special requirements on user interfaces to provide accessibility to individuals with disabilities. "In 1998, Congress amended the Rehabilitation Act of 1973 (29 U.S.C. 794d) as amended by the Workforce Investment Act of 1998 (P.L. 105 – 220), to require Federal agencies to make their electronic and information technology (EIT) accessible to people with disabilities. Specifically, Section 508 of that act requires that when Federal agencies develop, procure, maintain, or use EIT, Federal employees with disabilities have access to and use of information and data that is comparable to the access and use by Federal employees who are not individuals with disabilities, unless an undue burden would be imposed on the agency." [Section508 2011]

Expectations With Respect To Deliverables – Project deliverables should clearly document user interfaces. Where standards have been used, it should be so indicated. Where appropriate, Section 508 rules must be clearly implemented.

Recommendations and Guidance – Consideration should be given to the development of style guidelines for user interfaces for different types of homeland security M&S applications. Different types of M&S applications will undoubtedly require different interfaces, e.g., analytical versus training models and simulations. Such guidelines would be useful to developers and help create consistency across applications that would facilitate and simplify their use.

For Further Information – [Schaffer 2004] provides an overview on how to develop a graphical user interface standard. [Smith 1986] provides detailed information on the user interface design process, terminology, etc., that was developed for the U.S. Air Force.

4.3   Input and Output Data Types

Introduction – Models and simulations for homeland security applications typically will involve diverse sets of input and output data. Currently, simulation analysts have not only the burden of creating complex, technically-correct M&S applications, but often they also have to create the data sets needed to exercise those applications. If data sets already exist, they may not be in usable form. Analysts may have to create translators, import, and export software to obtain the data from other HS applications. Examples of the types of data sets that may be needed to support homeland security simulations follow (some data types may appear in multiple categories):

- *Incidents*: incident summaries, chronologies, timing descriptions, victims, damage assessments, response operations, status, models, message logs, media files, reports and other records, and after action reviews.
- *Incident management structure and resources*: organizations including non-governmental organizations (NGOs), roles, responsibilities, policies, plans, standard operating procedures (SOPs), actions, records, resource allocations, checklists, facilities, equipment, systems, vehicles, personnel, evacuation centers, supplies, documentation media, contact points, data, capabilities, resource capacity, status, and funds.
- *Infrastructure systems*: transportation (roads, airports and airlines, trains, buses, trucks, privately-owned vehicles, etc.), telecommunications, power, gas, water, food, healthcare, sewage, alerting systems, status, and sensitive targets.
- *Spatial data*: various maps, navigation charts, terrain models, definitions of geographical regions, areas, building layouts, models of structures, layout of pipelines, schematics of electrical distribution systems, locations of highways, rail lines, and locations of other critical infrastructure systems.
- *Environment data*: climate, weather (precipitation, wind speed, air temperature, etc.), societal, political, economic, biosphere, and chemical properties/hazard effects data.
- *Hazard effects*: chemical, biological, nuclear, fire, severe weather, and other natural and man-made disasters; plume models; and health.
- *Responder computing and communications*: radio and other equipment, channel assignments, switching systems, transmission towers, areas of coverage, and message formats.
- *Demographics and Behavioral Data*: population characteristics, social behavioral data for those populations, location, age, sex, and other attributes of population by time of day.
- *Financial*: cost of operations, consumables, leased equipment, and labor.
- *Controlling Documents*: policies, plans, protocols, and procedures.

- *Investigative Intelligence*: crime scene forensics as well as various databases that are mined to gather intelligence for combating terrorism including locations, facilities, organizations, individuals, components, documents, money, weapons, vehicles, and drugs (see [Mena 2004] for an in depth analysis of this data type).
- *Training data*: course syllabi, lesson plans, instructional materials, tests, exercises, and references.
- *Systems Engineering data*: requirements analyses, system design specifications, system documentation, test plans and procedures, and test data sets.
- *Simulation Support data*: software/game assets, statistical distributions, and programming scripts.

NIST staff has identified a number of standards that exist for homeland security M&S applications that are the focus of this report. One of the objectives of a recent joint DHS-NIST M&S for Homeland Security Workshop focused on review of existing data standards. [NIST 2011b] provides a summary report of the workshop.

Terminology:

>**Taxonomy** – is a classification scheme, typically a hierarchical structure, for organizing and categorizing objects according to certain attributes of those objects.

>**Information model** – is a representation of concepts, relationships, constraints, rules, and operations to specify data semantics for a chosen domain of discourse. It can provide stable, sharable, and organized structure of information requirements for the domain context [Wiki 2011e].

>**Data dictionary** – "A data dictionary, or meta-data repository is a centralized repository of information about data such as meaning, relationships to other data, origin, usage, and format. The term may have one of several closely related meanings pertaining to databases and database management systems (DBMS):

>- a document describing a database or collection of databases
>- an integral component of a DBMS that is required to determine its structure
>- a piece of middleware that extends or supplants the native data dictionary of a DBMS." [Wiki 2011f]

>**Parser** – is a computer program that reads input in the form of source program instructions, commands, markup tags, or some other defined data interface. The parser typically breaks up data by data type, tags the data types, and checks that the syntax (structure) of data types is correct. The parsed data file is then typically passed to other programs for further processing, e.g., program compilation.

Issues:

>**Potential Problem Area(s)** – Simulation analysts will have a number of obstacles to overcome with respect to data. Does data exist in a computer-interpretable format? For example, during a disaster data capture is not often a high priority; thus good historical data on past incidents may not exist. Analysts may be denied access to data for security, privacy, or political reasons. Is the data valid? Data may exist but do we know that it is correct? If data exists and is correct, is it in a usable digital format? Do common keys exist that can be used to cross-reference related data types? The lack of standard formats for many data remains a long-term issue that needs to be

addressed. Finally, policies need to be set with respect to which data formats should be used by homeland security M&S applications.

Relevant Technologies – Information modeling languages can be used to analyze data requirements and define structures for homeland security data types. Database management systems can be used to efficiently manage large and small data sets alike. Parsers, software compilers, and data consistency checkers can be used to verify that data files are formatted correctly according to the specifications.

Possible Role of Standards – Although a number of additional standards are needed for information models and data formats, many already exist. The appendices to [NIST 2011b] identify many data standards that already exist, e.g., Emergency Data Exchange Language (EDXL). A number of standards that can be used to develop new data standards are also available. The Unified Modeling Language (UML) can be used to develop information models for different data types. Extensible Markup Language (XML) can be used to define the formats of data messages and files. The Interface Definition Language (IDL) can be used to "describe the interfaces that client objects call and that object implementations provide. An IDL interface definition fully specifies each operation's parameters and provides the information needed to develop client applications that use the interface's operations." [Webopedia 2011a]

Expectations With Respect To Deliverables – Documentation deliverables associated with M&S applications should clearly identify: data types used, definitions of those data types, sources of data, how data was validated, standard formats and languages used to store and access data, identification of keys used to cross-reference different data types for linked data sets, information on how to modify existing data sets or create new ones, and versions/configuration management data. If there are any security or privacy issues associated with data, they should also be identified.

Recommendations and Guidance – Policies and guidelines should be established that identify the different types of data that are needed for the modeling and simulation of different types of applications. Applications include M&S of incidents, critical infrastructure systems, training, and exercises. The policies should also identify a suite of recommended data standards for different interrelated data types. A data dictionary that defines data types and a repository of validated reference data sets should be established. The dictionary and repository would facilitate the development of new M&S applications by simulation analysts. It would help enable compatibility and re-use of M&S applications created by different developers.

For Further Information – For a summary of existing data standards that may be used by M&S applications see [NIST 2011b]. [DHS 2011a] provides information on a taxonomy of critical infrastructure system data that is currently under development by DHS.

4.4   Databases, Data Files, and Translators

Introduction – The primary formats for storing data used by homeland security M&S applications are data files and databases. The databases and data files are information stored external to the simulation application. Databases will typically be implemented using relational or object-oriented technologies. Relational databases are essentially collections of tables. Translators will often be required to reformat data maintained in files or databases into data structures and data types that are usable by the simulation environment. Output data from the simulation may also need to be translated, if external data files or databases are to be updated.

Database management systems (DBMS) typically have a data dictionary that defines the structure of data contained within its databases. Data files defined using the Extensible Markup Language (XML) will contain embedded tags and data type definitions that can be used to automate the interpretation of data. Other standard legacy data file formats may have their structures defined externally to the file. In this case, custom software may have to be generated to read and write these data files.

Terminology:

> **Structured Query Language (SQL)** – a standard computer language that is used to access and update information stored in a relational database.

> **Extensible Markup Language (XML)** – a standard computer language that is used to tag and structure data contained within a document file so that it can be automatically read by a computer program.

Issues:

> **Potential Problem Area(s)** – Similar types of data used by different organizations that have a stake in homeland security may have vastly different structures and data elements. Models and simulations will typically be designed to process data in one consistent format. As such, M&S software that is widely used may require the development of many custom translators to convert information into the format of a given simulator.

> A related problem has to do with cross-referencing data contained in different files or databases. The term *key(s)* is given to a data element that is used to extract a particular data item or set of items from a database. For example, a social security number is a unique key often used to extract personnel data from a database. First and last name is a compound key that may not be unique, as multiple individuals may have the same name. Homeland security databases created by different organizations will not necessarily use the same or even compatible keys, making it difficult or impossible to cross-reference data.

> **Relevant Technologies** – Software tools such as compilers, or YACC and Lex found in the Unix operating system, can be used to create translators to automatically restructure data for use by other programs. Numerous high-end as well as low-cost database-management systems provide tools for defining and manipulating data structures, as well as accessing and updating data. There are a number of commercially available tools for working with XML.

> **Possible Role of Standards** – XML and SQL are two standards that provide functionality to access files and databases that should satisfy most near-term homeland security M&S needs.

> **Expectations With Respect To Deliverables** – Project deliverables should identify and provide samples of databases and data files used in the implementation of the application. Software documentation should provide instructions on how to modify queries or data files access routines/translators so that the application can be updated as data or formats change.

Recommendations and Guidance – Facilities that will run homeland security M&S applications should identify data file formats and database management systems that are used in their facility. If both low-end and high-end DBMS systems are available, they should be so identified.

For Further Information – For information on the XML standard, see [DuCharme 1999]. For an online tutorial on SQL, see [SQLTutorial 2011]. For information on the use of YACC and LEX to create data translators, see [Brown 1992].

4.5   Information Security Mechanisms

Introduction – Given the potential sensitive nature of homeland security data, protection of that data from viewing by unauthorized personnel can be critical. Sensitive data may include responder resources, response plans, healthcare data, and CI vulnerabilities.

Unfortunately most, if not all, COTS simulation environments and public domain simulation software are designed without security features in mind. Simulation software does not contain embedded security mechanisms to protect information. The only security mechanisms that are typically included within COTS software are designed to prevent unauthorized use of unlicensed copies of the software.

Simulations that are run within the protection of secure DHS facilities are not necessarily a problem. Although in such facilities, it may still be possible for users to access data through a simulation that they are not normally authorized to view if appropriate precautions are not taken. Also distributed simulations that have been integrated using the High Level Architecture (HLA) may use communication security mechanisms contained within specific implementations of a given Run Time Infrastructure (RTI). These RTI mechanisms may determine which data a specific application may be allowed to view.

A significant problem may arise if homeland security M&S applications are deployed for use by state and local governments and other organizations. Organizations outside of DHS may or may not have adequate security measures in place to protect sensitive information from local databases or prevent hacking of IT systems.

[Gobuty 1998] presents information on performing security certifications and accreditations for simulator/simulation facilities. A typical security accreditation document might contain:

- Prescribed security operating mode.
- Accredited sensitivity level.
- Data allowed to be stored, processed, and protected.
- Specified physical environment, internal, and external connections.
- Approved operating procedures, users, and operators.
- Times and dates that it may operate.
- Mission the system may complete.
- Residual security risks that approval authority agrees to assume.

A key element of information security is access control. Access controls may typically be either identity-based (i.e., based on an authenticated user identity) or rule-based (i.e., based on policies for specific users or types of users). The complex nature of simulations may complicate access control. A simulation may require access to many different types of sensitive data to function. Some or all users of the simulation may not be granted access to that sensitive data, i.e., it will be invisible to them within the context of the simulation. Developers will need to carefully structure user interfaces and implement access controls to ensure that sensitive data is not viewed or compromised by unauthorized personnel.

Terminology:

Information security – "… means protecting information and information systems from unauthorized access, use, disclosure, disruption, modification, or destruction in order to provide:

(A) integrity, which means guarding against improper information modification or destruction, and includes ensuring information non-repudiation and authenticity;
(B) confidentiality, which means preserving authorized restrictions on access and disclosure, including means for protecting personal privacy and proprietary information;
(C) availability, which means ensuring timely and reliable access to and use of information."
[USCode 2011]

Issues:

Potential Problem Area(s) – State and local governmental agencies, non-governmental organizations (NGOs), academia, and other private institutions (e.g., healthcare facilities) that may ultimately run HS M&S applications (e.g., for training purposes) may not have necessary expertise or capabilities to adequately evaluate and ensure information security. Sensitive data that may be compromised may be contained in local databases.

Relevant Technologies – A number of tools are commercially available to help plan, implement, and evaluate information security. One such tool, the Security Assessment Simulation Toolkit (SAST), provides a suite of simulation tools directly applicable to cyber security training, exercises, testing, evaluation, information assurance, and information operations [Meitzler 2009].

A number of software products are available to encrypt data, control access to data, and authenticate users.

Possible Role of Standards –NIST publishes a number of standards relating to information security as part of its Federal Information Processing Standards (FIPS) publication series. These standards deal with topics such as security requirements, user authentication, local area network security, passwords, digital signatures, and data encryption. These standards may be useful in the establishment of information security mechanisms for homeland security M&S applications.

Expectations With Respect To Deliverables – M&S project documentation should clearly identify if sensitive data is involved or may be added by future application users. Security mechanisms that have been implemented should be clearly identified.

Recommendations and Guidance – DHS should identify potential information security vulnerabilities associated with M&S applications and solutions for addressing those vulnerabilities.

For Further Information – For further information of Federal Information Processing Standards (FIPS) publication series, see [FIPS 2011]. An overview of information security policies, architectures, processes, certification, and accreditation of secure systems that focuses on DoD simulation applications can be found in [Gobuty1998]. Information on DoD simulation security policy can be found in [DoD 2006b].

5 The M&S Development Process

Simulation textbooks typically recommend that a ten to twelve step process be followed in the development of models and simulations. The recommended M&S development approach from [Banks 1996] usually involves the following steps:

1. Problem formulation.
2. Setting of objectives and overall project plan.
3. Model conceptualization.
4. Data collection.
5. Model translation into computerized format.
6. Code verification.
7. Model validation.
8. Design of experiments to be run.
9. Runs and analysis.
10. Documentation and reporting.
11. Implementation.

See also [Banks 1998], [Kelton 1998], or [Law 2000] for further information on this approach.

Unfortunately, the approach described above often leaves the simulation analyst with considerable work and possibly too much creative responsibility. The analyst is often faced with answering very broad, open-ended questions, such as:

- What is the problem that needs to be simulated?
- What are the objectives of the simulation?
- What is the conceptual model of the problem?
- How will the real world system be abstracted for implementation within the simulator?
- What data is available and how will it be collected?
- What simulation system or language will be used?
- What input and output data formats are required?
- Will the simulator need to interact with other application such as other software and database?
- Do new data formats need to be developed?
- What probability distributions are needed to approximate the behavior of the real system?
- How will the model be verified and validated?
- How many times should the model be run to ensure statistically significant results?
- What conclusions can be drawn from the simulation results?
- How will the simulation be deployed?
- What documentation will be required?

Using the "textbook" approach, at its current level of specificity, the process of modeling and simulation is perhaps as much an art as it is a science. Simulations are often developed from scratch, so the skill of the individual analyst may figure significantly in the quality and interpretation of the results that are obtained. With current simulation technology, there is little opportunity for the analyst to build upon the work of others since each simulation is built as a custom solution to a uniquely defined problem. Input data from other software applications is not often in the format required for simulation, so data must often be abstracted, reformatted, or translated. Furthermore, pressure from management to obtain quick results may have an adverse effect on the performance of the simulation analyst and the quality of results obtained.

In addition to those that have been identified above, large modeling and simulation projects raise a number of other key issues that need to be addressed, namely:

- Project team and developer qualifications.
- Analysis of user needs and system requirements.
- Simulation specifications: models, designs, and interfaces.
- Software risk management.
- Proprietary commercial versus open source software.
- Use of standards.
- Distributed simulation architectures and communications.
- Verification, validation, accreditation, and testing.
- Documentation and training.
- Training.

These issues are briefly addressed in the subsections that follow.

5.1   Project Team and Developer Qualifications

Introduction – One of the most critical issues that needs to be addressed in the review of any homeland security M&S proposal is the project team and developer qualifications. Who will be members of the development team? Do the members of the proposed team have the necessary skills and experience to effectively implement the proposed M&S application(s)?

[Taylor 2003] provides a table that identifies possible players and roles in the conduct of a simulation study. Taylor's work is based upon an earlier paper addressing more generic research teams, see [Omerod 2001]:

Doers
- Project manager – Responsible for managing the process; may not have specific modeling skills
- Modeler – Develops the model (conceptual and computer)
- Model user (in later stages) – Experiments with the model to obtain understanding and look for solutions to the real world problem

Done for
- Clients – The problem owner and recipient of the results directly or indirectly funds the work
- Model user (in early stages) – Recipient of the model

Done with
- Data providers – Subject matter experts who are able to provide data and information for the project
- Modeling supporter – A third party expert (software vendor, consultant or in-house expert) provides software support and/or modeling expertise

Done to
- Those interviewed for information – A wide group of people from whom information is obtained

Done without
- Management, staff, customers – Beneficiaries of the project, but not involved; in some cases they are not aware of the project. [Taylor 2003]

Of all the potential participants in a project team, the skills and abilities of the simulation analyst are perhaps most critical for a traditional M&S project. In [Rohrer 1998], the skills required of a simulation analyst are described. The tasks performed by a simulation analyst include data collection, conceptual model development, specification development, model construction, verification, validation, experimentation, and presentation of results. A paraphrased summary of each of the simulation analyst skills and tasks identified by Rohrer follows:

- Data collection – involves gathering information on the system to be modeled, performing time studies, reviewing spreadsheet or text files, analyzing input data for accuracy, fitting distributions to data, acquisition control algorithms and rules used to operate a system.
- Conceptual model development – a conceptual model must be created before any specification documents are written or model development begins – this phase, also called model formulation, is defined by [Shannon 1975] as the "Reduction or abstraction of the real system to a logic flow diagram."
- Specification development – specification development is the description of the conceptual model in words that are written for a non-simulation audience so that all individuals involved in the project understand how the model will be constructed. It summarizes the input data and approach used to build the model.
- Model construction – translation of the conceptual model into a computer model within the framework of the simulation package which may include some programming, using CAD-like capability, process flow development, and interaction through the software's user interface. Verification – Verification is defined as a comparison of the conceptual model with the computer model. There are many methods of verification.
- Validation – Validation is determining whether the model reflects reality. It is extremely important since the basis of simulation is the substitution of the computer model for the real system. Experimentation – Experimentation is conducting "what-if" analyses with the model, the analyst gains understanding of the behavior of the system under varying conditions. Organizational skills to keep track of scenarios and results; analytical skills; and statistical competency are required to understand techniques such as design of experiments, warm-up determination, and confidence intervals.
- Reports and presentations – Reporting results is the technical documentation of the experimentation outcome. It may include making inferences from the output data. Skills in report writing, technical presentations, and persuasion are key to successfully communicating model results.
- Other necessary skills include listening, negotiating, a strong work ethic, and the ability to build rapport with customers. [Rohrer 1998]

In many fields (e.g., medicine, engineering, and accounting), aside from simulation, there are professional certifications that provide a means for nominally evaluating a potential team members credentials, i.e., domain experts. For simulation analysts, this evaluation may be more difficult. There currently are no widely recognized certification processes for simulation analysts that have stood the test of time. Recently efforts have been undertaken address this deficiency through the creating of a simulation analyst certification process by the SimSummit Consortium [SimSummit 2011] and National Defense Industrial Association (NDIA) [MSPCC 2011].

The development of game-based simulations will require additional skills beyond those required to create traditional simulations, see [Bethke 2003] and [Flynt 2005] for a discussion of game development team software engineering roles and special skill requirements (e.g., story writers; 3D/2D graphics artificial intelligence, audio, physics, and network communications programmers; artists and animators; and various modelers.)

Terminology:

> **Professional certification** – "a designation earned by a person which shows that s/he is qualified to perform a job or task. Professional certification is also known as trade certification or professional designation. Certifications are usually earned from a professional society or educational institute, and not the government. In the U.S., professional certificates are issued by state agencies. Professional certificates are valid for a specific period of time and must be renewed periodically. Professional certifications are common in areas such as health care, aviation, construction, technology, and other industrial sectors." [USLegal 2011]

Issues:

> **Potential Problem Area(s)** – Development teams may lack relevant skills or expertise in model development or in relevant technical domains. Some technical areas may lack adequate professional certification programs that could be used to evaluate team qualifications. Even with certification, abilities of analysts to solve different types of problems may vary widely. After a project has been awarded, team members may change due to normal turnover or other reasons.

> **Relevant Technologies** – There are a number of team building tools that have been developed by various management consultants and professional organizations that can help formulate and improve the team building process.

> **Possible Role of Standards** – Professional certifications represent a form of personnel qualification standards. Relevant professional certifications in different homeland security, engineering, and other technical domains should be identified and used wherever possible as a factor in evaluating project team qualifications. Academic degrees in relevant areas should also be used in the qualification evaluation process.

> **Expectations With Respect To Deliverables** – Project proposals should clearly identify members of the project team, their credentials, e.g., academic degrees, certifications, technical training, and professional experience. Proposals should also address risk management for personnel and how turnover of employees will be addressed.

Recommendations and Guidance – Managers of homeland security projects in the different technical domains, i.e., CIKR, IM, HS, and HMR, should create a reference document that identifies the types of experts associated with each HS domain. This information might be placed in tables that also identify expected academic credentials, professional certifications, and experiential qualifications that are appropriate to each domain. This information would be helpful to COTRs in the evaluation of future project proposals.

For Further Information – The World Health Organization has prepared an excellent white paper on technical team building [WHO 2007].

5.2   Analysis of User Needs and System Requirements

Introduction – Perhaps the most critical step in the development process is the analysis of user needs and the specification of system requirements. Texts on this phase often use varying or conflicting terminology for needs and requirements. For purposes of this document, user-needs analysis is the identification of high-level needs that must be met to satisfy the user or customer for a system. System-requirements specifications translate user needs into specific functional or other capabilities that must be implemented in the system to satisfy those needs.

With respect to needs analysis, [Wieringa 1996], states "… the function that the system must have for the user is determined. Needs analysis establishes why a system should exist. It is the problem analysis task of the engineering cycle. It is oriented to the environment of the product, i.e., to the client or to the market. It produces a specification of the objectives that the system must satisfy."

[Sodhi 1992] describes the information gathering process that is typically associated with defining user needs: "fact-finding, validating the customer's understanding of gathered information, communicating open issues for resolution. Fact-finding uses mechanisms such as interviews, questionnaires, and observation of the operational environment of which the software will become a part. Validation involves creating a representation of elicitation results in a form that will focus attention on open issues that can be reviewed with those that provided the information. Possible representations include summary documents, usage scenarios, prototypes, and graphic models…."

After user needs are fully captured and understood, it is the objective of the requirements specification phase to translate those needs into system requirements.

[Sommerville 2004] identifies four notations for capturing and specifying requirements, namely: 1) structured natural language, 2) design description languages, 3) graphical notations, and 4) mathematical specifications. He also defines how the software requirements document is used by different personnel:

- System customers – Specify the requirements and read them to check that they meet their needs. Customers specify changes to requirements.
- Managers – Use the requirements document to plan a bid for the system and plan the system development process.
- System engineers – Use the requirements to understand what system is to be developed.
- System test engineers – Use the requirements to develop validation tests for the system.
- System maintenance engineers – Use the requirements to understand the system and relationship between its parts. [Sommerville 2004]

The IEEE STD 830-1998 Software Requirements Specification [IEEE 1998] provides a template for document development. A general description of contents follows:

- *Introduction*
  - *Purpose*
  - *Scope*
  - *Definitions, acronyms, abbreviations*
  - *References*
  - *Overview*
- *Overall Description*
  - *Product Perspective*
    - *System interfaces*
    - *User interfaces*
    - *Hardware interfaces*
    - *Software interfaces*

- *Communications interfaces*
- *Memory*
- *Operations*
- *Site adaptation requirements*
- *Product Functions*
- *User Characteristics*
- *Constraints*
- *Assumptions and Dependencies*
- *Apportioning of requirements*
- *Specific Requirements*
  - *External interfaces*
  - *Functions*
  - *Performance requirements*
  - *Logical database requirement*
  - *Design constraints*
  - *Key features*
- *Appendices*
- *Index*

The complete text of the specification is available from [IEEE 1998].

Terminology: The IEEE Standard Glossary of Software Engineering Terminology [IEEE 1990] defines many terms relating to the software engineering process:

Requirement – "(1) A condition or capability needed by a user to solve a problem or achieve an objective. (2) A condition or capability that must be met or possessed by a system or system component to satisfy a contract, standard, specification, or other formally imposed documents. (3) A documented representation of a condition or capability as in (1) or (2)." [IEEE 1990]

Issues:

Potential Problem Area(s) – M&S projects often either: 1) devote inadequate time and resources to needs analysis and requirements specification, 2) possibly skip this phase altogether, or 3) indicate that they will produce this documentation once the target system has been completed. In the third case, one can compare this to a builder who says "Go ahead sign the contract – don't worry about the specifications, I'll prepare the specifications for your house after its completed." Results of this needs analysis/requirements specification phase should ultimately drive the modeling, design, and testing phases of system development. Without a clear definition of requirements, testing and verification become essentially meaningless activities.

Relevant Technologies – There are many analytical and diagramming techniques that may be used to perform needs analysis and requirements specifications. See "For further information" below for various texts on this subject.

Possible Role of Standards – A number of standard formats for requirements specifications have been defined by the standards development organizations (SDOs), e.g., IEEE, ISO.

Expectations With Respect To Deliverables – A document that specifies user needs and system requirements should be an essential deliverable for all M&S projects. The document should be completed and reviewed by the sponsors before further work is initiated on the project.

Recommendations and Guidance – For all M&S projects, DHS managers/COTRs should ensure that users/customers of systems or systems outputs have been identified before development begins. Users of M&S application outputs may not necessarily come in direct contact with the system. Analysis should clearly define user needs and specify system requirements arising from those needs. Needs analyses and system requirements specifications should be clearly documented and reviewed by the customer organization before the green light on modeling, system design, and development begins. Traceability matrices should be created to map system design features and tests against system requirements.

For Further Information – For in depth texts on needs analysis and requirements gathering techniques, specification processes, diagramming techniques, document templates, see [Donaldson 1997], [Sodhi 1992], [Davis 1993], [Macaulay 1996], [Wieringa 1996], and [Caputo 1998].

5.3 Simulation Specifications: Models, Designs, and Interfaces

Introduction – Once user needs are captured and simulation requirements are specified, conceptual models, detailed designs, and interface specifications for the simulation must be developed. Models are typically simplified representations of reality. Models are created to better understand complex systems that currently exist or are under development. Four aims of models are to 1) visualize an existing or proposed system, 2) specify the structure or behavior of a system, 3) serve as a template for guiding the construction a system, and 4) document system design decisions [Booch 1998].

Models are used to [Rumbaugh 1999]:

- *Capture and precisely state requirements and domain knowledge so that all stakeholders may understand and agree on them...*
- *Think about the design of a system...*
- *Capture design decisions in a mutable form separate from the requirements...*
- *Generate usable work products...*
- *Organize, find, filter, retrieve, examine, and edit information about large systems...*
- *Explore multiple solutions economically...*
- *Master complex systems...*

The Unified Modeling Language (UML) is a powerful tool for development of conceptual models and system designs for homeland security simulations. UML is widely used in the software community. It has become the modeling tool of choice for a large number of software as well as other engineering projects. UML provides a number of solutions for graphically specifying aspects of a system including requirements, architectures, module interactions and functionality, system designs, data structures, behaviors, interfaces, tests, etc. [Booch 1999] introduces five interlocking UML views that can be used to describe software intensive systems:

1. *Use case view* – describes the behavior of the system in terms of use cases of its end users, analysts, and testers.
2. *Design view* – supports the functional requirements of the system, i.e., services provided to end users in terms of object classes, interfaces, and collaborations between modules.
3. *Implementation view* – addresses the configuration management of system releases in terms of the component modules and files that comprise a version of the system.
4. *Process view* – addresses the performance, scalability, and throughput of the system in terms of the threads and processes that implement concurrency and synchronization mechanisms.
5. *Deployment view* – addresses the distribution, delivery, and installation of the physical system components on the nodes that form the hardware topology on which the system executes.

A number of UML diagrams have been defined to represent system models and designs. These diagrams are briefly defined in the terminology section below.

Once system models and designs are developed, interfaces must be specified. Typically interfaces may fall into the following categories: file formats, messages, subroutine calls (application programmer interfaces), and database queries on shared databases. Interface Definition Language (IDL) and Structured Query Language (SQL) provide standards for the specification of subroutine calls on code associated with objects and access to relational databases. (See Section 5.3, Input and Output Data Types, for further information on these topics.)

Extensible Markup Language (XML) provides tools for the structured information interchange and the specification of interfaces. A number of homeland security data standards have been developed using XML, e.g., Emergency Data Exchange Language (EDXL) and Common Alerting Protocol (CAP). XML defines a set of rules for encoding information in machine-interpretable form, i.e., it is a markup language. Some of the key components of XML are tags, data elements, attributes, document type definitions, and schemas. A detailed discussion of XML is beyond the scope of this document.

Terminology: UML diagram definitions excerpted from [Booch 1998]:

> **Class diagram** – *shows a set of classes, interfaces, collaborations, and their relationships* ...

> **Object diagram** – *shows a set of objects and their relationships* ...

> **Use case diagram** – *shows a set of use cases and actors (a special kind of class) and their relationships* ...

> **Sequence diagram** – *an interaction diagram that emphasizes the time-ordering of messages* ...

> **Collaboration diagram** – *an interaction diagram that emphasizes the structural organization of objects that send and receive messages* ...

> **State chart diagram** – *shows a state machine, consisting of states, transitions, events, and activities* ...

> **Activity diagram** – *is a special kind of state chart diagram that shows the flow from activity to activity within a system* ...

> **Component diagram** – **shows** *the organizations and dependencies among a set of components* ...

> **Deployment diagram** – *shows the configuration of run-time processing nodes and the components that live on them* ...

Issues:

> **Potential Problem Area(s)** – Potential developers/simulation analysts may need to be motivated or contractually required to develop appropriate conceptual models and simulation designs before actual coding of simulations begins. Simulation analysts often would rather jump right into coding (programming) than perform up front modeling and design activities.

Staff of development organizations will require training in UML modeling techniques and tools to make proper use of this technology. Appropriate time and funds should be allocated to training to ensure staff is properly prepared during the conceptual modeling/design phases of the project.

Appropriate UML modeling tools will need to be acquired by development organizations. A number of licenses may be required depending upon the size and composition of the development team.

Relevant Technologies: The UML and associated modeling/Computer-Aided Software Engineering (CASE) tools provide required functionality to perform conceptual modeling and simulation design for DHS homeland security M&S applications. The tools will generate appropriate reports, diagrams, and may automatically generate software code.

Possible Role of Standards: UML is in widespread use today for modeling large and small software, as well as other engineering projects. The Object Management Group (OMG) has standardized UML. It is an appropriate standardized solution to address DHS needs in this area.

Expectations With Respect To Deliverables: A set of hierarchical UML conceptual models and system designs should be included in the delivered documentation set for virtually all homeland security M&S projects. The type, number, and detail of the UML diagrams should be appropriate to the size and complexity of the simulation that is being developed.

Recommendations and Guidance – DHS should establish policies regarding the development of conceptual models, designs, and interface specifications for homeland security simulations. UML is a modeling language with broad functionality that has stood the test of time and should meet DHS needs. Depending on the nature and size of the project, UML models could be a contractual deliverable. Modeling guidance might also be developed that suggests the types of UML diagrams and level of detail required based on the system size, complexity, expected number of users, and deployment plans for a simulation.

A publicly accessible library of UML models (for models that are not sensitive) might also be established to facilitate re-use of models by developers. Such a model library should help the simulation community develop a common understanding of homeland security systems, resources, operations, processes, and data.

XML could provide a standard mechanism for the definition of homeland-security data file formats and interfaces. Any new standard data file or message formats should probably be based on XML technology.

A common approach to the use of UML and XML may help facilitate the development of compatible simulations as well as needed standards.

For Further Information – There are a number of excellent texts available on UML modeling, e.g., see [Alhir 1998], [Booch 1998], and [Rumbaugh 1999]. For specific examples on how to model enterprises that may be helpful in modeling CI and IM systems, see [Marshall 2000]. For information on the design and engineering of systems, decision tools (various types of models, simulations, and tradeoff analyses), see [Kossiakoff 2003]. See [Goldfarb 2001] for an overview of XML, detailed explanations of features, and tutorials. For detailed documentation on the XML specification, see [DuCharme 1999]. See [Grady 1994] for a step-by-step guide to the interface development process including interface analysis and schematic methods.

5.4   Software Risk Management

Introduction – Software projects are notorious for scheduling delays, cost overruns, or failures to meet technical expectations or performance requirements. Within the context of this report, most simulation development activities are software projects. As such, simulation development is subject to the risks associated with software projects. [Sommerville 2004] describes a software risk management process that includes risk identification, risk analysis, risk planning, and risk monitoring. Examples of different types of software management risks suggested include:

- Technology – derived from technologies associated with development software and hardware.
- People – issues associated with the development team, such as recruiting appropriate staff, staff availability, and lack of training.
- Organizational – risks due to management changes, financial problems, and budget changes.
- Tools – development tools and support software does not perform as expected, e.g., computer-aided software engineering (CASE) tools or database management systems.
- Requirements – problems associated with changes to customer requirements that require significant design changes, or management of the requirement change process.
- Estimation – estimates of system size/complexity are incorrect, inadequate resources allocated for development efforts, schedules are overly optimistic, and bugs take longer than expected to fix.

Terminology: Definitions of steps of the risk management process from [Sommerville 2004] include:

    Risk Identification – Possible project, product, and business risks are identified.

    Risk Analysis – The likelihood and consequences of these risks are assessed.

    Risk Planning – Plans to address the risk either by avoiding it or minimizing its effects on the project are drawn up.

    Risk Monitoring – The risk is constantly assessed and plans for risk management are revised, as more information about the risk becomes available.

Issues:

    Potential Problem Area(s) – Simulation system vendors may not be in the habit of developing software risk management plans or sharing information contained within these plans with potential customers. Given the opportunity, it is likely that many potential vendors may attempt to avoid this topic.

    Relevant Technologies: The Capability Maturity Model (CMM) developed by the Software Engineering Institute for the Department of Defense provides a number of tools for improving the software development process as well as tracking and managing risks.

    Expectations With Respect To Deliverables: A software risk management plan should be included a part of project proposals. Vendors should provide periodic status reports and updates on risks and risk management efforts.

    Possible Role of Standards: "ISO 31000:2009 sets out principles, a framework and a process for the management of risk that are applicable to any type of organization in public or private sector. It does not mandate a "one size fits all" approach, but rather emphasizes the fact that the management of risk must be tailored to the specific needs and structure of the particular organization ... At the same time, ISO is publishing "ISO Guide 73:2009, *Risk Management Vocabulary*," which complements ISO 31000 by providing a collection of terms and definitions relating to the management of risk. " [ISO 2009]

Recommendations and Guidance – The level of effort and specificity of a risk management plan should be driven by and appropriate to the simulation project involved. Projects that involve large systems, significant costs, widely distributed simulations, critical schedules, or systems where human life or safety may be involved, should give more emphasis to risk management. Development proposals should address risk management. Proposal reviews should evaluate and weigh risk management plans accordingly.

For Further Information – See [Sommerville 2004] for a detailed discussion of the software risk management process, examples, and techniques for reducing risk. For more information of the Software Engineering Institute (SEI) Capability Maturity Model and tools that are available to help management the software development process, see [Caputo 1998]. For current work on the development of international standards for risk management, see [ISO 2009].

5.5   Proprietary Commercial versus Open Source Software

Introduction – M&S development projects will typically involve the creation of new software using either simulation languages, simulator programming extensions (e.g., macro programming capabilities), or programming languages (e.g., C/C++ code or other general purpose programming languages). An M&S application will typically be comprised of both proprietary licensed commercial simulators, as well as homeland security application code (programs) for which under the government should presumably own the intellectual property rights (IPR). Contracts should be clear on these issues to avoid confusion and possibly additional development or distribution costs at a later date.

A major issue that will need to be addressed with each M&S project is intellectual property rights (IPRs) associated with the M&S code, i.e., who really owns the code? IPR will determine whether DHS has access to source code, if changes can be made (e.g., scenario changes and game modifications), who may make changes, whether software can be copied, how many copies can be made, and whether those copies may be distributed to other agencies/organizations.

Although many simulation analysts may be able to create M&S applications through the use of commercially-developed simulation tools without resorting to low-level programming, in some cases developers will need to create low-level code, e.g., C or C++ code. Use of existing commercial software has a number of advantages in terms of maturity, support, and cost. On the down side, the cost of distributing licenses to commercial tools may be prohibitively expensive. Some simulators will allow developers to compile distribution versions of simulations that recipients cannot modify, but with significantly reduced licensing costs. Also, if developers/analysts need to write low-level code to develop M&S applications, there may be more problems with compatibility with existing systems, reliability, and security.

Access to M&S source code will help ensure that the homeland security users will be able to make necessary changes to simulation software as technology evolves; or missions, scenarios, operations, and resources change. Open source code should be created and used, wherever appropriate and possible. Some of the advantages of open source code are that other developers can see the code and make enhancements, there is less duplication of effort through the re-creation of the same code over and over, and increased collaboration opportunities to work on shared code, rather than proprietary products.

A NASA Technical Report provides excellent background and guidance on the issue of open source code; for more information, see [NASA 2003]. Some of the key points from the report are excerpted below:

> Open source refers to idea that the source code to an application is provided along with the executable code....With open source code, there is the potential for improved software quality, more efficient software development, and increased collaboration. An organization known as the Open Source Initiative (OSI) provides the most widely recognized guidelines as to what constitutes Open Source; in particular the OSI provides guidance with respect to how to balance the intellectual property rights concerns of developers with openness... The appeal of Open Source software for users is based on more than simply the desire to avoid paying money or to make a political statement....Software that is available Open Source is easier to evaluate before making a deeper commitment....In cases of closed or proprietary software, there may be no recourse for the user. Open Source software, on the other hand, leaves open the option of stepping in to keep the software alive. Enhanced collaboration would tend to produce software with more users, and in particular users who have a vested interest in seeing the software continue to thrive. In the long term those users may see value added in a commercialization that provides systems integration and support (think, for example, Linux). Enhanced dissemination would also tend to create a larger user base, and a large user base would enable more commercial opportunities. In the long term it may be worthwhile for the government to address software distribution, and in particular Open Source, in the initial request for proposals and statement of work. [NASA 2003] (For further information on OSI, see [OSI 2011b].)

In 2000, the President's Information Technology Advisory Committee (PITAC) was convened to address the subject of "Developing Open Source Software to Advance High End Computing," one of the committees' recommendations was:

> The Federal Government should allow open source development efforts to compete on a 'level playing field' with proprietary solutions in government procurement of high end computing software. Requests for Proposals (RFPs) from Federal agencies for high end computing software, tools, and libraries should include provisions allowing these efforts to be carried out using open source. [PITAC 2000]

Terminology:

> **Open source code** – "Denoting software for which the original source code is made freely available and may be redistributed with or without modification." [Google 2011] The Open Source Initiative [OSI 2011a] further expands the definition, addressing topics such as:
>
> 1) *Free Redistribution – The license shall not restrict any party from selling or giving away the software as a component of an aggregate software distribution containing programs from several different sources. The license shall not require a royalty or other fee for such sale...*
> 2) *Source Code – The program must include source code, and must allow distribution in source code as well as compiled form. Where some form of a product is not distributed with source code, there must be a well-publicized means of obtaining the source code for no more than a reasonable reproduction cost preferably, downloading via the Internet without charge. The source code must be the preferred form in which a programmer would modify the program. Deliberately obfuscated source code is not allowed. Intermediate forms such as the output of a preprocessor or translator are not allowed...*

3) *Derived Works – The license must allow modifications and derived works, and must allow them to be distributed under the same terms as the license of the original software ...*

4) *Integrity of The Author's Source Code – The license may restrict source-code from being distributed in modified form only if the license allows the distribution of "patch files" with the source code for the purpose of modifying the program at build time. The license must explicitly permit distribution of software built from modified source code. The license may require derived works to carry a different name or version number from the original software...*

5) *No Discrimination against Persons or Groups – The license must not discriminate against any person or group of persons...*

6) *No Discrimination Against Fields of Endeavor – The license must not restrict anyone from making use of the program in a specific field of endeavor...*

7) *Distribution of License – The rights attached to the program must apply to all to whom the program is redistributed without the need for execution of an additional license by those parties...*

8) *License Must Not Be Specific to a Product – The rights attached to the program must not depend on the program's being part of a particular software distribution. If the program is extracted from that distribution and used or distributed within the terms of the program's license, all parties to whom the program is redistributed should have the same rights as those that are granted in conjunction with the original software distribution...*

9) *License Must Not Restrict Other Software – The license must not place restrictions on other software that is distributed along with the licensed software. For example, the license must not insist that all other programs distributed on the same medium must be open-source software...*

10) *License Must Be Technology-Neutral – No provision of the license may be predicated on any individual technology or style of interface...*

Open architecture – "An architecture whose specifications are public. This includes officially approved standards as well as privately designed architectures whose specifications are made public by the designers. The opposite of open is closed or proprietary. The great advantage of open architectures is that anyone can design add-on products for it. By making an architecture public, however, a manufacturer allows others to duplicate its product. Linux, for example, is considered open architecture because its source code is available to the public for free. In contrast, DOS, Windows, and the Macintosh architecture and operating system have been predominantly closed." [Webopedia 2011b]

Issues:

Potential Problem Area(s) –
- Costs associated with obtaining the number of licenses needed for M&S development, maintenance, and distribution using proprietary commercial software.
- Conversely, costs associated with developing, maintaining, and supporting users of open or government owned source code.
- Issues associated with the soundness/longevity of commercial software developers – how long will the developers be in business?
- Problems with keeping legacy software operational as developers continue to modify underlying M&S software and issue new releases of that software.
- Costs associated with maintaining current licenses of underlying M&S software.

- Problems dealing with proliferation of versions of homeland security applications and underlying M&S software that are widely distributed.

Relevant Technologies – SourceForge.net provides a means for developing and using software that is voluntarily developed and maintained by a vast community of developers. IPR for SourceForge software contains special licensing restrictions that keep code open to future users/developers.

Possible Role of Standards – An open architecture clearly defines modules and interface points by which systems may be interconnected. The establishment of standard open architectures for different types of M&S applications would help ensure that legacy software would be compatible with future software development. Standards efforts for open architecture M&S applications do not currently exist.

Expectations With Respect To Deliverables – Development proposals should clearly identify all software (by name, developer/owner, license type) required to develop, use, distribute, and maintain homeland security M&S applications.

Recommendations and Guidance – Topics raised in this section may have a major impact on the functionality, availability, and costs associated with homeland security M&S applications over the long term. Efforts spent educating stakeholders on these issues and establishing appropriate policies are likely to reap significant benefits for DHS. M&S software development and IPR policies must be based on a number of criteria including possibly software availability, estimated development costs, technology ownership (patents/copyrights), intended use, projected life span, and number of users/installation sites.

For Further Information – For further information on open source code development via SourceForge, see [SourceForge 2011]. For more about the Open Source Initiative, open software architectures, etc., see [OSI 2011b].

5.6   Use of Standards

Introduction – Widespread use of standards could help the homeland security community make more effective and efficient use of M&S by enabling reuse of models, sharing of data, integration of independently developed modules, etc. In Information Technology Standards: Quest for the Common Byte, Martin Libicki explains the role of standards in information systems:

> Standards solve particular problems, such as how to represent data efficiently or manage a communications system, and they create benefits—interoperability, portability, ease of use, expanded choice, and economies of scale—that only exist when many systems do things the same way. [Libicki 1995]

Standards should support the design, development, and implementation of the M&S applications. Examples of major categories of standards that may be relevant to these homeland security M&S applications include:

- Architectures.
- Integration interfaces
  - General purpose integration interfaces.
  - Domain-specific integration interfaces.
- Equipment specifications.

- Operational guidelines.
- Document formats.

Architectures support the overall design or structure of a system or system environment. *Integration Interface* standards facilitate the interoperation or data exchange between systems. *General Purpose Integration Interfaces* are used to integrate a wide variety of computer applications and are not specific to homeland security or related mission areas. Example interfaces include markup languages, image file formats, and database query languages. *Domain-specific Integration Interfaces* are specific to homeland security-related areas, e.g., emergency communications message formats. *Equipment Specifications* define required capabilities, functional characteristics, or rules that ensure quality, safety, and health of users. *Operational Guidelines* define organizational structures, policies, procedures, and protocols. *Document Formats* specify layout and structure for documents in word processing, database, spreadsheet, graphic, presentation, printed and encoded formats.

A joint DHS-NIST report on the 2011 Modeling and Simulation for Homeland Security Workshop, identified a large number of standards that may be relevant to the implementation of homeland security M&S applications [NIST 2011b]. Standards that are relevant to specific types of homeland security M&S applications are categorized by the four domains identified in Section 5.3 of this document. Generic standards that are applicable to multiple areas are also grouped into separate categories in this document. The document does not recommend which standards should be adopted by DHS to meet specific needs, it merely indicates standards that are currently available by various categories. The development of a recommended suite of standards was beyond the resources and schedule of the workshop.

Terminology: Standards-related definitions from [NRC 1995]:

> **Standard** – *A prescribed set of conditions or requirements concerning definition of terms; specification of performance, operation, or construction; delineation of procedures; or measure of quantity and/or quality in describing features of products, processes, systems, interfaces, or materials.*

> **Voluntary consensus standard** – *A standard arising from a formal, coordinated process in which key participants in a market seek consensus. Use of the resulting standard is voluntary. Key participants may include not only designers and producers, but also consumers, corporate and government purchasing officials, and regulatory authorities.*

> **Conformity assessment** – *The determination of whether a product or process conforms to particular standards or specifications. Activities associated with conformity assessment may include testing, certification, accreditation, and quality assurance system registration.*

> **Certification** – *The process of providing assurance that a product or service conforms to one or more standards or specifications. Some, but not all, certification programs require that an accredited laboratory performs any require testing.*

> **Accreditation** – *The process of evaluating testing facilities for competence to perform specific tests using standard test methods.*

Issues:

> **Potential Problem Area(s)** – The standards development process is typically time-consuming. The development of a single standard often takes several years. Given tight budgets and limited

resources, getting the appropriate individuals and organizations involved and committed to the standards development process may be difficult. Some organizations will argue that development of standards may hurt their competitive advantage by opening the market to competing and compatible products.

Relevant Technologies: UML and XML are two technologies that have been demonstrated to be very valuable to the standards process. These two technologies are widely used for the development of interface specifications and standards for information technology (IT) systems. Many software tools are currently available that implement these technologies and facilitate their use. See conceptual modeling and interface specifications in Section 5.3, "Simulation Specifications: Models, Designs, and Interfaces," for further discussions of these topics.

Expectations With Respect To Deliverables – M&S applications should clearly identify the standards that have been used to implement simulations. Where certification-testing capabilities are available, M&S applications should be tested to ensure that they have correctly implemented appropriate standards.

Recommendations and Guidance – Appropriate organizations within DHS, e.g., programs acquiring M&S applications, should work with the homeland security M&S user and development community to set policies for selection and use of standards in each technical domain for which applications are being developed. Particular emphasis should be placed on data and interface standards. Gaps should be identified where needed standards do not currently exist. If resources permit, efforts to develop standards to fill gaps should be identified and promoted within the voluntary consensus standards community, i.e., standards development organizations (SDOs) such as Simulation Interoperability Standards Organization (SISO), ISO, ANSI, or Institute of Electrical and Electronics Engineers (IEEE).

For Further Information – See [NRC 1995] for an in depth background on standards, their development, organizations involved, challenges, and recommendations. For more on the Unified Modeling Language (UML), see [Alhir 1998], [Booch 1998] and [Rumbaugh 1999]. For more on Extensible Markup Language (XML) see [Goldfarb 2002].

5.7   Distributed Simulation Architectures and Communications

Introduction – The simulation community has developed a number of different distributed simulation software architectures. The architectures are designed to enable simulation modules created by different developers to be combined into a single integrated simulation. What are the advantages of distributed simulation over monolithic, stand-alone simulations? A distributed approach to M&S software architectures could enable the integration of modules created by different developers and enhance overall functionality of analysis, planning, and training systems. A distributed approach to M&S can:

- Enable parallel, modular development of specialized simulation system components by independent software developers with different areas of expertise.
- Allow the configuration of integrated simulations that meet specific regional or scenario-based needs.
- Model multiple organizations where some of the information about the inner workings of each organization may be hidden from other participants for reasons such as security or proprietary issues.
- Simulate multiple levels of organizations and systems at different degrees of resolution such that lower level simulations generate information that feeds into higher levels.
- Model multiple systems with different simulation requirements where an individual simulation-vendor's products do not provide the capabilities to model all areas of interest.

- Allow software developers to hide the internal workings of a simulation system through the creation of run-time simulators with limited functionality.
- Create an array of low-cost, run-time simulation models that can be integrated into larger models.
- Take advantage of additional computing power, specific operating systems, or peripheral devices (e.g., virtual reality interfaces) afforded by distributing across multiple computer processors.
- Provide simultaneous access to executing simulation models for users in different locations (i.e., collaborative work environments.)
- Offer different types and numbers of software licenses for different functions supporting simulation activities (e.g., model building, visualization, execution, and analysis.)

Two compatible architectures have been developed that are appropriate candidates to support the near-term implementation of distributed homeland security simulations, i.e., the DoD/IEEE High Level Architecture (HLA) and Sandia National Laboratories' SUMMIT architecture. But what is meant by architecture and why is it important?

> As the size and complexity of software systems increase, the design and specification of overall system structure become more significant issues than the choice of algorithms and data structures of computation. Structural issues include the organization of a system as a composition of components; global control structures; the protocols for communication, synchronization, and data access; the assignment of functionality to design elements; the composition of design elements; physical distribution; scaling and performance; dimensions of evolution; and selection among design alternatives. [Shaw 1996]

DoD and the Simulation Interoperability Standards Organization (SISO) initiated the development of the HLA, but it is now an IEEE standard for implementing distributed simulation. The HLA architecture is also known as the IEEE 1516 standard. In HLA terms, individual simulations are called federates and the distributed simulation is referred to as a federation. The HLA defines a framework by which individually executing federates can be combined into a distributed simulation federation.

The HLA framework has three major parts. The first part is a set of rules that federates and federations must adhere to ensure that a federation operates properly. The second part is a software system called the Run-Time Infrastructure (RTI). The RTI defines an interface that provides a number of services that federates can use to communicate (i.e., exchange simulation data), and coordinate their execution (i.e., synchronize simulation clocks) with other federates in a federation. The third part of the HLA is called the Object Model Template (OMT). The OMT provides a means for describing the format of the data that will be exchanged between federates. The IEEE standards for each of these three major components of HLA are briefly described below:

> *Framework and Rules – IEEE Standard P1516:* The HLA rules describe the responsibilities of federates (simulations, supporting utilities, or interfaces to live systems) and federations (sets of federates working together to support distributed applications). The rules comprise a set of underlying technical principles for the HLA. For federations, the rules address the requirement for a federation object model (FOM), object ownership and representation, and data exchange. For federates, the rules require a simulation object model (SOM), time management in accordance with the HLA Runtime Infrastructure (RTI) time management services, and certain required functionality and constraints on attribute ownership and updates.

> *Federate Interface Specification – IEEE Standard P1516.1:* In the HLA, federates interact with an RTI (analogous to a special-purpose-distributed operating system) to establish and

maintain a federation and to support efficient information exchange among simulations and other federates. The HLA interface specification defines the nature of these interactions, which are arranged into sets of basic RTI services.

Object Model Template (OMT) Specification – IEEE Standard P1516.2: The HLA requires simulations (and other federates) and federations to each have an object model describing the entities, not necessarily platform entities, represented in the simulations and the data to be exchanged across the federation. The HLA object model template prescribes the method for recording the information in the object models, to include objects, attributes, interactions, and parameters, but it does not define the specific data (e.g., vehicles, unit types) that will appear in the object models.

Other HLA standards include IEEE 1516.3 – Recommended Practice for High Level Architecture Federation Development and Execution Process (FEDEP) and IEEE 1516.4 – Recommended Practice for Verification, Validation, and Accreditation of a Federation an Overlay to the High Level Architecture Federation Development and Execution Process.

The Standard Unified Modeling, Mapping, and Integration Toolkit (SUMMIT) is a software toolkit that allows analysts, emergency planners, responders, and decision makers to seamlessly access integrated suites of modeling tools and data sources for planning, exercises, or operational response. SUMMIT enables:

- Discovery and use of models, simulations, data, and archived analyses that are relevant to a specific scenario or event of interest
- Bringing together users and modeling resources from many locations while ensuring that access to existing data and models is controlled by the resource owners
- Automatically linking together disparate modeling tools
- Playing "what if" scenarios for complex incidents
- Archiving and managing analysis results (including model configuration parameters)
- Use of novel visualization and collaboration environments for enhancing user understanding of simulation results [SUMMIT 2011]

Sandia National Laboratories, the developers of the SUMMIT architecture, state that

SUMMIT is not intended as an alternative to HLA; it operates at a higher level by identifying appropriate resources and relying on frameworks like HLA for the actual integration. There are several levels at which resources must be integrated. These include low-level communication, semantic compatibility, and graphical interface aggregation … Central to our approach is the notion of a federation execution architecture (FEA). We partition the set of all possible simulations into FEAs that are directly integrable based on intent or technology. For example, all simulations built using a particular HLA Runtime Infrastructure (RTI) would comprise an HLA FEA. Within an FEA, many simulations can be directly connected, while others can be easily adapted. This use of direct connections means that simulations will be efficiently usable by the SUMMIT system without special considerations by their developers. [Friedman-Hill 2010]

Terminology:

Federate – is an independently executing simulation with in HLA federation.

Federation – is a collection of HLA simulation federates, which are integrated by a Federation Object Model and services provided by a Run-Time Infrastructure.

Simulation Object Model (SOM) – specifies the shared object, attributes, and interactions of a single HLA federate.

Federation Object Model (FOM) – specifies the shared object, attributes, and interactions for an entire HLA federation.

Object Model Template (OMT) – defines the framework for the communication between HLA federates and is comprised of FOM and SOM documents.

Run-Time Infrastructure (RTI) – is software that manages the execution of an HLA federation and handles communications between processes.

Issues:

Potential Problem Area(s) – Current HLA standards address a number of integration issues, but they are not complete integration specifications. The quick and reliable integration of individually developed simulations is not necessarily a simple or straightforward process. The development of an HLA federation requires additional specification work beyond what is contained in the standard, e.g., SOMs, the FOM, and selection of RTI services that are used. The additional specifications will require agreement on the part of all of the developers that are contributing a federate(s) to a given HLA federation.

A number of different RTIs have been implemented and are commercially available. Federates implemented to work with a specific RTI cannot be expected to function with another RTI without additional work.

Some potential users of HLA, e.g., the gaming community, have been hesitant to adopt HLA due to issues associated with the overhead of implementing HLA federations and the performance of RTIs and HLA federations.

Relevant Technologies: A number of HLA RTIs and associated development tools are commercially available from both U.S. as well as overseas developers. These tools should meet DHS needs for building distributed simulations.

Possible Role of Standards: HLA is an IEEE standard with considerable industry and government agency support. The SUMMIT architecture, while not technically a standard, is designed to facilitate the integration of simulations using the HLA architecture. The National Exercise Simulation Center (NESC) is committed to using simulations developed according to the SUMMIT architecture.

Expectations With Respect To Deliverables – For projects implementing distributed simulations or modules that may one day be integrated into a distributed simulation, HLA and SUMMIT technologies should be employed wherever possible. Deliverables should include appropriate documentation on HLA FOMs, SOMs, RTI software versions, RTI service calls, and SUMMIT capabilities that were used to implement the simulation.

Recommendations and Guidance – DHS should establish policies regarding the use of distributed simulation architectures. It should also review RTIs and indicate approved RTIs for distributed simulation implementations. If resources permit, DHS should coordinate the specification of SOMs and FOMs that can be used by the simulation community to develop distributed homeland security simulations, as well as establishing test data sets for those SOMs and FOMs.

For Further Information – For information on the development of distributed simulations using HLA as well as example software exercises, see [Kuhl 2000]. The text also contains a disk with software that can be used to develop simple HLA federations for training purposes. A number of other distributed simulation architectures, associated specifications, and standards have been developed (e.g., Distributed Interactive Simulation (DIS), Test and Training Enabling Architecture (TENA), Common Training and Instrumentation Architecture (CTIA), and Aggregate-Level Simulation Protocol (ALSP)), for more information on these other architectures, see [Richbourg 2008].

5.8   Verification, Validation, Accreditation, and Testing

Introduction – Evaluation of M&S capabilities and results should take into account the many factors that affect the quality of the results. Factors include the level of understanding or knowledge of the issues being addressed and the experience level of model developers. Various forms of testing, verification, validation, and accreditation provide a means for assuring the quality of models and simulations.

Verification and testing are closely related. Verification evaluates the accuracy of the implementation of a model. Testing is one type of capability used to verify models and simulations (example of another verification technique that does not use testing is a walkthrough.) Some forms of software testing may be used for verification purposes while some may satisfy other needs. Some examples of different types of testing that may be relevant to a typical M&S project include unit testing, integration testing, acceptance testing, alpha-beta testing, and conformance testing. For definitions of these different types of testing, see the terminology section below.

Several organizations have developed processes and guidelines to address the credibility of M&S capabilities. For example, DoD has a process for documentation, evaluation, and certification of M&S results known as Verification, Validation and Accreditation (VV&A) that is defined in a recommended practice guide [DoD 2006a]. The DoD process is implemented in policy, which develops a common understanding of the major steps in the VV&A process [DoD 2009].

All M&S capabilities should complete verification and validation (V&V). M&S capabilities used as the primary input to critical decision making, e.g., on cost, schedule, or performance of the system, should be formally accredited to certify that the results are credible for their intended use. Analysts and decision-makers need to be aware of these sorts of issues when presented with computational modeling and simulation results. [Balci 1998] developed a taxonomy and describes the use of different methods to evaluate M&S capabilities and results based on software testing approaches.

While [Sterman 2000] argues the ability to validate models, the book does provide excellent insights into what can be done in the area of VV&A and what one can conclude from VV&A activities.

Terminology:

Unit testing– individual units or modules of source code are tested to determine whether they are suitable for use.

Integration testing – after units or modules successfully pass unit testing, they are grouped or aggregated for combined testing.

Acceptance testing – the test of a system to ensure that contractual requirements are satisfied.

Alpha-beta testing – internal-alpha or external-beta (limited marketplace release) testing of a software product to find bugs/needed improvements before final release.

Conformance testing – testing to determine whether or not a product conforms, i.e., correctly implements a formal standard.

Test plan – "a management planning document that shows: 1) How the testing will be done (including SUT (system under test) configurations), 2) Who will do it, 3) What will be tested, 4) How long it will take (although this may vary, depending upon resource availability), and 5) What the test coverage will be, i.e. what quality level is required." [Wiki 2011g, based upon the IEEE 829 standard]

Test case – a set of data used to implement a specific test.

Test suite – a group of test procedures that are used to evaluate specific functionality or behaviors of a system under test.

Verification – The process of determining that a model implementation and its associated data accurately represents the developer's conceptual description and specifications [DoD 2009].

Validation – The process of determining the degree to which a model and its associated data are an accurate representation of the real world from the perspective of the intended uses of the model [DoD 2009].

Accreditation –The official certification that a model, simulation, or federation of models and simulations and its associated data are acceptable for use for a specific purpose [DoD 2009].

Issues:

Potential Problem Area(s) – There is considerable controversy on how to perform VV&A for simulation models. With limited project resources, VV&A is an activity that may be dropped in favor of other development efforts. Without adequate VV&A, it may not be clear exactly what an M&S application has modeled and whether the results drawn from that model can be trusted.

Relevant Technologies – There are many VV&A techniques available, see [Balci 1998] for a summary of those techniques. Software tools are commercially available to automate testing processes, so as to minimize the need for human testers to perform many interactions with a system under test.

Possible Role of Standards –SISO is about to launch an effort in V&V. The SISO GM-VV Product Development Group (PDG) – Generic Methodology for Verification and Validation "proposes to develop a product for the international community for a generic V&V and Acceptance methodology for models, simulation, and data." [SISO 2011] The standard will include:

- *The Handbook, which safely guides its users through the V&V and Acceptance efforts and clarifies their responsibilities by explaining how to apply the methodology in practice. It describes the activities to perform and the products to produce, the interactions taking place among those involved, the flow of products, and how to tailor the methodology to the specific needs of the Modeling and Simulation (M&S) project.*

- *The Reference Manual documents the underlying concepts of the methodology, including the foundations of the chosen terminology, the explanation of the dependencies between activities and products, their meaning for the V&V and Acceptance endeavor, and the rationale for their execution and creation. The reference manual is referred to whenever a deeper understanding of the methodology is required.*

- *The Recommended Practices document provides user specific guidance with regards to the selection and use of techniques and tools in support of the Handbook. This will include domain specific case studies thereby illustrating the application and tailoring of the methodology.*

The products stemming from this PN (Product Nomination) will serve all communities managing, developing, and/or using M&S and in particular those charged with the task of conducting and managing V&V and Acceptance activities. The community spans many different user domains (e.g. Training, Analysis, and Acquisition) and application areas (e.g., Defense, Entertainment, Medical, Space). [SISO 2011]

Mil-Std-3022, Department Of Defense Standard Practice: Documentation Of Verification, Validation, And Accreditation (VV&A) For Models And Simulations (28 Jan 2008) establishes templates for the four core products of the Modeling and Simulation (M&S) Verification, Validation, and Accreditation (VV&A) processes: Accreditation Plan; V&V Plan; V&V Report; and Accreditation Report. Specific templates for each document are found in the appendices [EverySpec 2011].

IEEE 829-2008 Standard for Software Test Documentation defines test plan terminology and provides a standard format for developing software test plans.

Other IEEE and ISO standards exist for software quality assurance that may useful to the development of homeland security M&S testing, VV&A policies, and procedures.

Expectations With Respect To Deliverables – Supporting VV&A documentation should be a required deliverable of all homeland security M&S projects. The documentation should clearly identify the procedures used to test, verify, and validate models/data. Test data sets should be provided with delivered models that demonstrate capabilities that were called out in requirements specifications documents. Without this information, it will be very difficult to determine what, if any, confidence may be placed in an M&S project's results.

Recommendations and Guidance – DHS should use existing VV&A standards and policies where available due to the cost of developing and maintaining new standards. DHS Director of Standards can assist with the development of new policies and standards. At present, there are no commercial standards for V&V of generic M&S capabilities. DHS could benefit from a review of existing VV&A policies/procedures that have been established by other organizations, drawing on them wherever appropriate. Establishment of test data sets for different homeland security M&S domains could help developers evaluate models and simulations.

For Further Information – See [Balci 1998] for a summary of M&S VV&A techniques. Unfortunately, the summary does not discuss M&S uncertainties. For information on the differentiation between the assessment of a simulation and the verification and validation of general software products, see [Knepell

1993]. This text provides systematic, procedural, practical guides that can be used to enhance the credibility of simulation models. It includes assessment procedures and phases, discusses ways to tailor the methodology for specific situations and objectives, and provides numerous assessment aids.

For a general introduction to topics relating to software testing, see [Sommerville 2004], [Beizer 1984], or [Hetzel 1998]. For guidelines on software test planning and formats for test plan documents, see [Donaldson 1997]. For testing techniques that are specific to system dynamics models and simulations, see [Sterman 2000]. For discrete event model and simulation VV&A techniques see [Balci 1998].

For further information on VV&A policies, procedures, and activities other government agencies and organizations, the following references that may be useful:

- NASA developed a standard for M&S that is aimed at system development and includes V&V. The standard tailors the V&V approach to a given application depending on whether or not the results support decision making for mission critical or safety of flight issue. The standard also considers evaluation of the uncertainty of simulation results [NASA 2008].
- The American Institute of Aeronautics and Astronautics (AIAA) and the American Society of Mechanical Engineers (ASME) have both developed V&V guidelines/standards to evaluate computational fluid dynamics and computational solid mechanics models, respectively [AIAA 1998].
- Sandia National Laboratory developed guidelines on V&V to support nuclear weapons engineering and stockpile certification. V&V experts at Sandia claim the DoD VV&A process is necessary but not sufficient to ensure proper evaluation of various types of capabilities. The DoD VV&A process indicates what to do to perform VV&A for all types of M&S capabilities (analysis, engineering, training, cost estimating, etc.), but does not provide guidance on how to implement the process for different types of M&S capabilities. For example, performing VV&A on finite element models to address computational science and engineering problems would use different methods and tools compared to VV&A for training simulations [Sandia 1998].
- See [Oberkampf 2010] for a recent publication on "Verification and Validation in Scientific Computing," which is an excellent reference for evaluating computational science and engineering M&S capabilities. See also [Pace 2004] for an overview of challenges in the verification and validation of models and simulations.

5.9   Documentation and Training

Introduction – Appropriate documentation and possibly training materials are needed to ensure the maximum utility and long term benefit of homeland security M&S applications. Without suitable documentation, many problems can arise. Without proper analysis and documentation of user needs and system requirements, an M&S application may not have the functionality it needs to satisfy the objectives for which it was constructed. Without conceptual model, system design, and interface specification documents, it will not be clear how a system will be implemented, perhaps requiring costly modifications later in the development process. Without suitable test plans, test cases and V&V documentation, it will be difficult to evaluate whether the system satisfies requirements, correctly implements the design, and is valid solution to the technical problems it is designed to address. Without high-quality user documentation, new users may find it difficult to install the system, use it for its intended purpose, or adapt it to new scenarios/data. Without source code documentation, it may be impossible to modify the software when future homeland security needs, database/data file formats, ancillary software applications, computer operating systems, or hardware platforms change.

Similarly, for systems that may be widely used, training courses or individual instructional materials may be necessary to educate users on the system.

The terminology section that follows briefly identifies some of different types of documentation that may be required deliverables for homeland security M&S applications. The list does not address internal documents that a developer may use part of its software engineering process, e.g., project plans, schedules, bug tracking, and internal reviews.

Terminology:

> **User needs analysis/requirements specifications** – analyze and identify the capabilities that the user needs/expects to see from the system; specify system requirements (functional, data, performance, user interface, etc.) that have been defined to satisfy user needs.

> **Conceptual models, designs, and interface specifications** – include providing diagrams and text description of models, design and allocation of functions to modules, data structure and interface definitions, system states, and sequences of operations.

> **Installation and setup instructions** – provide guidance or step-by-step instructions for the administrator or user on the computer installation, software configuration, and possibly integration with other applications. They also identify minimum system requirements, possible error conditions, and known bugs.

> **User manuals** – is documentation on how to use the system and perform certain functions, and may include walkthroughs of menus and data entry forms, examples, definitions of terms, screen captures, and handling of errors.

> **Test plans** – (see Section 5.8, Verification, Validation, Accreditation, and Testing)

> **VV&A documents** – (see Section 5.8, Verification, Validation, Accreditation, and Testing)

> **Source code documentation** – comments in the source code of an application including explanations of control logic, data formats, and variables definitions.

Issues:

> **Potential Problem Area(s)** – Software engineers, programmers, and simulation analysts almost universally would rather develop new software than prepare documentation for that software or for software that is already completed. Yet documentation is critical to ensure that the correct system is actually built, that others may use the system, and that the system can be maintained when changes are required. A common excuse that is heard it that documentation will be written once the code is completed. Often that never happens since for most coders programming tends to be fun, at least in comparison to technical writing.

> Another claim that may be made by simulation analysts is that documentation is not really necessary because only the analyst or members of the development team have the necessary skills or professional training to effectively use the model or simulation. For some types of highly technical models, there may be some truth to this claim. Regardless of these claims, documentation should be developed so that others may check their work and new staff can more quickly be brought up to speed on the models/simulations as turnover occurs.

For systems that are complex or that will potentially have a large number of users, the development training materials and courses may be appropriate. Staff that is involved in the creation of M&S applications may lack the necessary skills to develop training materials or conduct classes.

Relevant Technologies – Desktop word processing and diagramming tools provide most of the capabilities needed to develop good software documentation. Simple animation and presentation tools are available to develop course materials. Screen capture tools are also available that can be used to make animated videos of user inputs and system responses for computer-based or classroom training materials.

Possible Role of Standards – A number of standards for software documentation exist today. The SEI's Capability Maturity Model (CMM) provides tools for evaluating the maturity of software projects, in large part based upon software documentation. MIL-STD-498 established uniform requirements for software development and documentation. IEEE/EIA 12207.0 Standard for Information Technology – Software Life Cycle Processes officially replaced the military standard in 1998. ISO, International Electrotechnical Commission (IEC), and ANSI also have developed software documentation standards.

Expectations With Respect To Deliverables – The number and level of detail for M&S documentation should be appropriate to the size, complexity, and expected number of users of the system. Development of good documentation will consume time and resources, as such documentation requirements should be commensurate with the size of the project. The absolute minimum documentation that might be considered acceptable for even the smallest project would be a conceptual model/design document, a VV&A document, and user documentation. The conceptual model/design document could include a statement of user needs and system requirements, the conceptual model of the system, and description of its design. The VV&A document should describe how the model and simulation was verified and validated. Setup and installation instructions should be provided. The user documentation should explain not only how to run the system, but also how to adapt it for new scenarios or data. Training materials to support individual or classroom instruction may also be included as a project deliverable – especially if a large number of users are expected.

Recommendations and Guidance – In the absence of departmental policies for the documentation of M&S applications, it is likely that many projects will produce inadequate documentation. Software that does not have adequate documentation will have little long term, or even possibly short-term value. Criteria for determining the types of documents that are required may be based on the technical complexity of the model, technologies used to develop the model, development costs, expected useful lifetime, projected number of users, technical domain of the model, nature of test data, distribution plans, etc.

For Further Information – For further information on software documentation, see [Donaldson 1997], [Caputo 1998], and [Flynt 2005].

6 Conclusions

This document presented modeling and simulation (M&S) in a manner and at a level appropriate for DHS Program Managers (and other responsible or interested executives within and outside DHS) with significant experience and domain knowledge but with limited understanding of the M&S technology. The current status of M&S implementation in DHS was discussed including the relevant policies, DHS component organizations utilizing M&S, and DHS roles and responsibilities relevant to M&S. Homeland security domains that can primarily benefit from M&S applications were discussed including, critical infrastructure systems, incident management, hazardous material releases, and health care systems. Various simulation technologies, including system dynamics, discrete-event, agent-based, and physical-science-based were presented. Gaming systems and table-top exercise enabling technologies that utilize M&S for training applications were also discussed. An overview of development technologies that may be used by DHS contractors to build M&S-based systems was provided. These included simulation environments and languages that may be used across a range of homeland security applications, and learning content management systems and learning management systems that may be used for training applications.

Applications based on simulations typically employ a number of components. DHS Program Managers evaluating proposals for such applications need to have a good understanding of the typical primary components as do the executives with non-technical backgrounds for all the organizations involved. With this view, the components discussed in this document include the simulation engine; user interfaces; input and output data types; databases, data files, and translators; and information security mechanisms. DHS Program Managers and other interested executives should also look for a well-defined M&S development process in the proposals for simulation application. The major steps in M&S development process and associated topics were discussed. The steps included project team and developer qualifications, analysis of user needs and system requirements, conceptual modeling and simulation design, software risk management, proprietary commercial versus open source software, use of standards, distributed simulation architecture and communications, interface specifications, testing, verification, validation, and accreditation (VV&A), documentation, and training.

DHS Program Managers and other interested executives should find this document a very useful source of overview information on M&S for specifying and assessing proposals for simulation based applications and systems for homeland security. Poor and incomplete specifications of applications, in particular simulation-based applications, can lead to development of systems that frustrate the users and create bad impressions of M&S technology in general. Perhaps to a lesser extent but still significantly damaging may be selection of contractor proposals for development of simulation-based applications without a proper assessment of the technical aspects. Both of the preceding problems can happen with the recommended increase of M&S by DHS if there aren't enough DHS Program Managers and other executives available with the requisite understanding of M&S. This document is intended to prevent such problems. Wide adoption of the recommendations in this document by Program Managers and other interested executives should help DHS deploy successful simulation applications that in turn will help increase the use of M&S and in the long term help improve DHS's performance.

7 Selected Acronyms and Abbreviations

AAR/IP – After Action Report/Improvement Plan
ABS – Agent-Based Simulation
ADL – Advanced Distributed Learning Initiative
AHRQ – The Agency of Healthcare Research & Quality
AIAA – American Institute of Aeronautics and Astronautics
ALSP – Aggregate-Level Simulation Protocol
ANSI – American National Standards Institute
API – Application Programmer Interface

ASME – American Society of Mechanical Engineers
C/E – Controller and Evaluator Handbook
C2I – Center of Excellence in Command, Control and Interoperability
C4ISR – Command, Control, Communications, Computers, Intelligence, Surveillance, and Reconnaissance
CAMRA – Center for Advancing Microbial Risk Assessment
CAS – Course Authoring System
CASE – Computer-Aided Software Engineering
CBRNE – Chemical, Biological, Radiological, Nuclear, and Explosive
CDC – The Centers for Disease Control and Prevention
CGMOES – Coast Guard Maritime Operational Effectiveness Simulation
CI – Critical Infrastructure Systems
CIKR or CI/KR – Critical Infrastructure and Key Resources
CMM – Capability Maturity Model
COE – Center of Excellence
CONOPS – Concept of operations
COTR – Contracting Officer's Technical Representative
CPR – Cardiopulmonary resuscitation
CREATE – Center for Risk and Economic Analysis of Terrorism Events
CTIA – Common Training and Instrumentation Architecture
DBMS – Database Management System
DES – Discrete Event Simulation
DEVS – Discrete Events System Specification
DHS – Department of Homeland Security
DIS – Distributed Interactive Simulation
DMSO – Defense Modeling and Simulation Office
DoD – Department of Defense
DOE – Department of Energy and Design of Experiments
DOT – Department of Transportation
EDXL –Emergency Data Exchange Language
EEG – Exercise Evaluation Guides
EMS – Emergency Management System
EMT – Emergency Medical Technician
EOC – Emergency Operations Center
EPA – Environmental Protection Agency
ESF – Emergency Support Function
ExPlan – Exercise Plan
FAST – NISAC Fast Analysis and Simulation Team
FAZD – National Center for Foreign Animal and Zoonotic Disease Defense
FDA – Food and Drug Administration
FEA – Federation Execution Architecture
FEDEP – HLA Federation Development and Execution Process
FEMA – Federal Emergency Management Agency
FFRDC – Federally Funded Research and Development Center
FOM – Federation Object Model
GIS – Geographic Information System
GSP – General Simulation Program
GUI – Graphic User Interface
GWU – George Washington University
Hazmat – Hazardous Material
HAZUS-MH – Hazards U.S. Multi-Hazard

HHS – Health and Human Services
HITRAC – Homeland Infrastructure Threat and Risk Analysis Center
HLA – High Level Architecture
HMR – Hazardous Material Releases
HS – Healthcare Systems
HSEEP – Homeland Security Exercise and Evaluation Program
HSNRA –Homeland Security Nation Risk Assessment
HSPD – Homeland Security Presidential Directive
HSSAI – Homeland Security Studies and Analysis Institute
HSSTAC – Homeland Security Science and Technology Advisory Committee
HSTA – Homeland Security Threat Assessment
IASD – OIP Infrastructure Analysis and Strategy Division
I&A – Intelligence and Analysis
IC – Incident Command
IDL – Interface Definition Language
IEC – International Electrotechnical Commission
IED – Improvised Explosive Device
IEEE – Institute of Electrical and Electronics Engineers
IICD – Infrastructure Information Collection Division
IM – Incident Management
IMAAC – Interagency Modeling and Atmospheric Assessment Center
IND – Improvised Nuclear Device
INS – Incident of National Significance
ISO – International Organization for Standardization
IT – Information Technology
LCMS – Learning Content Management System
LLNL – Lawrence Livermore National Laboratory
LMS – Learning Management System
MIPS – Center for Maritime, Island and Port Security
MS or M&S – modeling and simulation
MSCO – DoD Modeling and Simulation Coordination Office
MSEL – Master Scenario Event List
MSWG – DoD Modeling and Simulation Working Group
NARAC – National Atmospheric Release Advisory Center
NASA – National Aeronautics and Space Administration
NCBSI – National Center for Border Security and Immigration
NCFPD – National Center for Food Protection and Defense
NDCIEM – Center for Natural Disasters, Coastal Infrastructure, and Emergency Management
NESC – National Exercise Simulation Center
NGB – National Guard Bureau
NGO – Non-Governmental Organization
NIBS – National Institute of Building Sciences
NIMS – National Incident Management System
NIPP – National Infrastructure Protection Plan
NISAC – National Infrastructure Simulation and Analysis Center
NIST – National Institute of Standards and Technology
NOAA – National Oceanic and Atmospheric Administration
NPS – National Planning Scenario
NRC – National Research Council
NRF – National Response Framework
NTSCOE – National Transportation Security Center of Excellence

OAD – Operations Analysis Division
OIP – Office of Infrastructure Protection
OMG – Object Management Group
OMT – Object Model Template
PACER – National Center for the Study of Preparedness and Catastrophic Event Response
PITAC – President's Information Technology Advisory Committee
PMO – Program Management Office
RAPID – Risk Assessment Process for Informed Decision Making
RDBMS – Relational Data Base Management System
RMA – Office of Risk Management & Analysis
RTI – Run-Time Infrastructure
S&T – DHS Science and Technology Directorate
S/L – State and Local
SCORM – Shareable Content Object Reference Model
SDO – Standards Development Organization
SEDI – Homeland Security Systems Engineering and Development Institute
SEI – Software Engineering Institute
SISO – Simulation Interoperability Standards Organization
SitMan –Situation Manual
SLX – Simulation Language with eXtensibility
SME – Subject Matter Expert
SOM – Simulation Object Module
SOP – Standard Operating Procedure
SSA – Sector-Specific Agency and Social Security Administration
SSP – Sector Specific Plans
START – National Consortium for the Study of Terrorism and Responses to Terrorism
SUMMIT – Standard Unified Modeling Mapping Integration Toolkit
T&E – Test and Evaluation
TENA – Test and Training Enabling Architecture
TTX – Table Top Exercise
UML – Unified Modeling Language
USCG – United States Coast Guard
USNORTHCOM – U.S. Northern Command
USPHS – U.S. Public Health Service
VBA – Visual Basic for Applications
VNN – Virtual News Network
VVA or VV&A – Verification, Validation and Accreditation
WMD – Weapons of Mass Destruction
XML – Extensible Markup Language

8 References

[ADLNET 2011]   "Advanced Distributed Learning Initiative – The Power of Global Collaboration." Available via: *http://www.adlnet.gov/*. (Accessed on 22 October 2011.)

[ADS 2011]   Annual Agent Directed Simulation Symposium, held as part of the Spring Simulation Conference. Available via: http://scs.org/?q=node/205 (Accessed on 30 October 2011.)

[AFCESA 2001]   Protective Actions For A Hazardous Material Release. Headquarters, Air Force Civil Engineer Support Agency, Tyndall Air Force Base, Florida 32403. Available via: http://orise.orau.gov/csepp/documents/planning/reports/misc-reports/HAZMAT.pdf (Accessed on 10 May 2012.)

[AIAA 1998]   "Guide for Verification and Validation of Computational Fluid Dynamics Simulations AIAA G-077-1998." Computational Fluid Dynamics Committee on Standards, American Institute of Aeronautics and Astronautics. January 1998.

[Alhir 1998]   Alhir, S.S. *UML in a Nutshell*. O'Reilly and Associates: Cambridge, MA. 1998.

[Anderson 2008]   Anderson, D.R. *Model Based Inference in the Life Sciences: A Primer on Evidence,* Springer Science+Business Media, LLC.: New York, NY. 2008.

[Axelrod 1997]   Axelrod, R. *The Complexity of Cooperation: Agent-Based Models of Competition and Collaboration*. Princeton University Press: Princeton, NJ. 1997.

[Balci 1998]   Balci, O. "Verification, Validation, and Testing." In *Handbook of Simulation: Principles, Methodology, Advances, Applications, and Practice.* Edited by J. Banks. Wiley-Interscience: New York, NY. 1998.

[Banks 1996]   Banks, J., J. S. Carson, B. L. Nelson. *Discrete Event Simulation.* Prentice-Hall: Upper Saddle River, NJ. 1996.

[Banks 1998]   Banks, J. (ed.). 1998. *Handbook of Simulation: principles, methodology, advances, applications, and practice.* New York: John Wiley and Sons. 1998.

[Banks 2010]   Banks, J., J. S. Carson II, B. L. Nelson, and D. M. Nicol. *Discrete Event System Simulation.* 5th ed. Prentice Hall: Upper Saddle River, NJ. 2010.

[Beizer 1984]   Beizer, B. *Software Testing and Quality Assurance*. Van Nostrand Rheinhold, New York, NY. 1984.

[Bethke 2003]   Bethke, E. *Game Development and Production*. Wordware Publishing: Plano, TX. 2003.

[Bonate 2006].   Peter L. Bonate, P.L., *Pharmacokinetic-Pharmacodynamic Modeling and Simulation.* Springer Science+Business Media, Inc: New York, NY. 2006.

[Booch 1998]   Booch, G., J. Rumbaugh, and I. Jacobsen. *The Unified Modeling Language User Guide: The Ultimate Tutorial to the UML from the Original Designers.* Addison-Wesley: Boston, MA. 1998.

[Brown 1992]   Brown, D., J.Levine, and T. Mason. *YACC and LEX.* O'Reilly Media: Sebastopol, CA. 1992.

[Caputo 1998]   Caputo, K. *CMM Implementation Guide: Choreographing Software Process Improvement.* Addison-Wesley: Boston, MA. 1998.

[CCL 2012]   "NetLogo." The Center for Connected Learning (CCL). Available via: http://ccl.northwestern.edu/netlogo/index.shtml (Accessed on 25 January 2012).

[Cloyd 2007]   Cloyd, E., A. P. Leonardi, D. I. Scheurer, and E. J. Turner. *"Establishing National Ocean Service Priorities for Estuarine, Coastal, and Ocean Modeling: Capabilities, Gaps, and Preliminary Prioritization Factors."* NOAA Technical Memorandum NOS NCCOS 57, Washington, DC, 2007.

[CSIS 2011]   "Simulations and Tabletop Exercises: Part of the: Homeland Security Archived Projects." Center for Strategic and International Studies. Available via: http://csis.org/program/simulations-and-tabletop-exercises (Accessed 22 November 2011.).

[Davis 1993]   Davis, A. *Software Requirements: Objects, Functions, and States.* Prentice-Hall: Upper Saddle River, NJ. 1993.

[DHS 2002a]   "Homeland Security Act of 2002." U.S. Department of Homeland Security. Available via: http://www.dhs.gov/xabout/laws/law_regulation_rule_0011.shtm (Accessed on 26 November 2011.)

[DHS 2002b]   "National Strategy for Homeland Security – July 2002." U.S. Department of Homeland Security. Available via: www.dhs.gov/xlibrary/assets/nat_strat_hls.pdf (Accessed on 26 November 2011).

[DHS 2007]   "The National Strategy for Homeland Security." Homeland Security Council. The White House. Washington, DC. October 2007. Available via: http://www.dhs.gov/xlibrary/assets/nat_strat_homelandsecurity_2007.pdf (Accessed on 26 November 2011)

[DHS 2008]   "National Incident Management System (NIMS)." U.S. Department of Homeland Security. Available via: http://www.fema.gov/emergency/nims/ (Accessed on 26 November 2011.).

[DHS 2011]   "DHS Acquisition Staffing Survey and Analysis Report, Acquisition Program." Management Division/Cost Analysis Division, Office of the Under Secretary for Management, U.S. Department of Homeland Security: Washington, DC. May 5, 2011.

[DHS 2011a]   "Infrastructure Data Taxonomy: Common Terminology for Describing Critical Infrastructure." Available via: http://www.dhs.gov/files/publications/gc_1226595934574.shtm (Accessed on 20 November 2011.)

[DoD 1995]   Department of Defense. Modeling and Simulation Master Plan, DoD 5000.59-P, 1995. Available via: http://www.dtic.mil/whs/directives/corres/pdf/500059p1.pdf (Accessed on 21 November 2011.)

[DoD 2007b]   U.S. Department of Defense, DoD Modeling and Simulation Management, DoD Directive 5000.59, Washington, DC, August 8, 2007. Available via: http://www.dtic.mil/whs/directives/corres/pdf/500059p.pdf (Accessed on 12 March 2012)

[DoD 2006a]   V"V&A Recommended Practices Guide." DoD Modeling and Simulation Coordination Office (MSCO). Available via: http://vva.msco.mil/Default.htm (Accessed on 26 November 2011.)

[DoD 2006b]   "Acquisition Modeling and Simulation Master Plan – April 17, 2006." Department of Defense. Available via: http://www.acq.osd.mil/se/docs/AMSMP_041706_FINAL2.pdf (Accessed on 26 November 201.1)

[DoD 2007a]   "Strategic Vision for DoD Modeling and Simulation." Office of the Director of Defense Research and Engineering, Washington, DC. August 24, 2007. Available via: http://www.msco.mil/files/Strategic_Vision_Goals.pdf (Accessed on 20 November 2011.)

[DoD 2007b]   U.S. Department of Defense, DoD Modeling and Simulation Management, DoD Directive 5000.59, Washington, DC, August 8, 2007. Available via: http://www.dtic.mil/whs/directives/corres/pdf/500059p.pdf (Accessed on 12 March 2012.)

[DoD 2009]   DoD Modeling and Simulation (M&S) Verification, Validation, and Accreditation (VV&A). Department of Defense Instruction Number 5000.61. December 9, 2009. Available via: http://www.dtic.mil/whs/directives/corres/pdf/500061p.pdf (Accessed on 21 November 2011.)

[Donaldson 1997]   Donaldson, S. and S. Siegel. *Cultivating Successful Software Development: A Practitioner's View*. Prentice Hall: Upper Saddle River, NJ. 1997.

[DoT 2011]   "Performance Measurement." Available via: http://www.ops.fhwa.dot.gov/perf_measurement/fundamentals/index.htm (Accessed on 20 November 2011.)

[DuCharme 1999]   DuCharme, B. *XML: The Annotated Specification*. Upper Saddle River, New Jersey: Prentice Hall. 1999.

[E-Learning 2011]   "About E-Learning: E-Learning Glossary." http://www.about-elearning.com/e-learning-glossary.html (Accessed on 12 November 2011.)

[Engquist 2009]   Engquist, B., P.Lötstedt, and O.Runborg (Editors). *Multiscale Modeling and Simulation in Science*. Springer-Verlag: Heidelberg, Germany. 2009.

[EOP 2007]   Executive Office of the President of the United States. "FEA Consolidation Reference Model Document, Version 2.3." October 2007, p. 56. Available via: http://www.whitehouse.gov/sites/default/files/omb/assets/fea_docs/FEA_CRM_v23_Final_Oct_2007_Revised.pdf (Accessed on 05 March 2012.)

[EPA 2002]   U.S. Environmental Protection Agency, Quality Assurance Project Plans for Modeling, EPA QA/G-5M. Office of Environmental Information, Washington, DC, 2002. Available via: http://www.epa.gov/quality/qs-docs/g5m-final.pdf (Accessed on 05 March 2012.)

[EPA 2009]   U.S. Environmental Protection Agency. "Guidance on the Development, Evaluation, and Application of Environmental Models." EPA/100/K-09/003, Washington, DC. March 2009.

[EverySpec 2011]   "MIL-STD-3022, DoD Standard Practice: Documentation Of Verification, Validation, And Accreditation (VV&A) For Models And Simulations." Available via: http://www.everyspec.com/MIL-STD/MIL-STD+(3000+-+9999)/MIL-STD-3022_4197/ (Accessed on 21 November 2011.)

[FEMA 2008]   "National Response Framework." Federal Emergency Management Agency (FEMA) Department of Homeland Security. January 2008. Available via: http://www.fema.gov/pdf/emergency/nrf/nrf-core.pdf (Accessed on 22 November 2011.)

[FEMA 2009]   "FEMA Fact Sheet: National Planning Scenarios." Federal Emergency Management Agency (FEMA). Available via: http://www.fema.gov/pdf/media/factsheets/2009/npd_natl_plan_scenario.pdf Accessed on 26 November 2011.)

[FEMA 2011]   Federal Emergency Management Agency (FEMA). NRF Resource Center. "Emergency Support Function Annexes." Available via: http://www.fema.gov/emergency/nrf/# (Accessed on 3 November 2011.)

[FHA 2011]   "Mass Prophylaxis." Florida Hospital Association. Available via: http://www.fha.org/acrobat/TCMassProp.pdf (Accessed 21 November 2011).

[FIPS 2011]   "FIPS Publications." Available via: http://csrc.nist.gov/publications/PubsFIPS.html (Accessed on 26 November 2011.)

[Flynt 2005]   Flynt, J.P. *Software Engineering for Game Developers*. Thomson Course Technology: Boston, MA. 2005.

[Free 2011]   "Vulnerability Analysis." The Free Dictionary. Available via: http://www.thefreedictionary.com/vulnerability+analysis (Accessed on 20

November 2011.)

[Friedman-Hill 2010]	Friedman-Hill, E., T. Plantenga, and H. Ammerlahn "Simulation Templates in the SUMMIT System." *Proceedings of the Spring Simulation Interoperability Workshop 2010.* Simulation Interoperability Standards Organization (SISO). 2010.
[GAO 1976]	U.S. Government Accountability Office. "Ways to Improve Management of Federally Funded Computerized Models, LCD-75-111." Washington, DC. 1976.
[GAO 1978]	U.S. Government Accountability Office. "Models and Their Role in GAO, PAD-78-84." Washington, DC. 1978.
[GAO 1979]	U.S. Government Accountability Office. "Guidelines for Model Evaluation. PAD-79-17." Washington, DC. 1979.
[GBRA 2010]	Hazard Mitigation Plan Update, Draft: September 2010. Guadalupe-Blanco River Authority. Available via: http://www.gbra.org/hazardmitigation/default.aspx (Accessed on 3 April 2011.)
[Gen 2008]	Gen, M., R.Cheng, L.Lin. *Network Models and Optimization: Multiobjective Genetic Algorithm Approach.* Springer-Verlag: London, England. 2008.
[Gobuty 1998]	Gobuty, D.E. "Information Security in Simulator Design and Operation." In Cloud. D. and L. Rainey (eds.). *Applied Modeling and Simulation: An Integrated Approach to Development and Operation.* McGraw-Hill: New York, NY. 1998. Pp. 579-596.
[Goldfarb 2001]	Goldfarb, C. and P.Prescod. *XML Handbook.* Prentice Hall: Upper Saddle River, New Jersey: 2001.
[Google 2011]	Google. " Open source." Available via: http://www.google.com/#hl=en&rlz=1W1ADRA_en&q=open-source&tbs=dfn:1&tbo=u&sa=X&ei=RW_JTrqjO-Pt0gG5sdAr&sqi=2&ved=0CB4QkQ4&bav=on.2,or.r_gc.r_pw.,cf.osb&fp=5117386cc9c78ce3&biw=1680&bih=849 (Accessed on 20 November 2011.)
[Grady 1994]	Grady, J. *System Integration.* CRC Press: Boca Raton, FL. 1994.
[Hetzel 1988]	Hetzel, B. *The Complete Guide to Software Testing.* QED Information Sciences, Wellesley, MA. 1988.
[Hollenbach 2009]	Hollenbach, J.W. "Inconsistency, Neglect, and Confusion; A Historical Review of DoD Distributed Simulation Architecture Policies" Paper 09S-SIW-077. Simulation Interoperability Workshop (SIW), San Diego, CA. March 23-27, 2009.
[HR 2007]	U.S. House of Representatives. "HR 487 – Recognizing the contribution of modeling and simulation technology to the security and prosperity of the United States, and recognizing modeling and simulation as a National Critical Technology." 110th Congress. June 14, 2007. Available via:

http://thomas.loc.gov/cgi-bin/query/D?c110:1:./temp/~c110L0CRdm::
(Accessed on 20 November 2011.)

[HSEEP 2011]   "About HSEEP." FEMA Homeland Security Exercise and Evaluation Program."
Available via: https://hseep.dhs.gov/pages/1001_About.aspx#Terminology
(Accessed on 20 November 2011.)

[HSPD 2011]   "Homeland Security Presidential Directives." U.S. Department of Homeland
Security. Available via: http://www.dhs.gov/xabout/laws/editorial_0607.shtm
(Accessed on 26 November 2011.)

[Hupert 2004]   Hupert N., J. Cuomo, M. Callahan, A. Mushlin, S. Morse. "Community-Based
Mass Prophylaxis: A Planning Guide for Public Health Preparedness –
Department of Public Health Pub No. 04-0044." Weill Medical College of
Cornell University, Agency for Healthcare Research and Quality (AHRQ)..
Department of Public Health, Rockville, MD: August 2004. Available via:
http://www.nhrmcurgentcare.org/documents/SERAC/Resources/cbmprophyl.pdf
(Accessed on 20 November 2011.)

[Hutchings 2010]   Hutchings, C.W. "Improving the Management of Modeling and Simulation
Capabilities in the U.S. Department of Homeland Security." U. S. Department of
Homeland Security, Science & Technology Directorate Workshop on Grand
Challenges in Modeling, Simulation and Analysis for Homeland Security
(MSHAS-2010), Arlington, VA, 17 – 18 March 2010.

[IEEE 1990]   "IEEE STD 610.12-1990 IEEE Standard Glossary of Software Engineering
Terminology." Institute of Electrical and Electronics Engineers. Available via:
http://ieeexplore.ieee.org/Xplore/login.jsp?url=http%3A%2F%2Fieeexplore.ieee
.org%2Fstamp%2Fstamp.jsp%3Ftp%3D%26arnumber%3D159342&authDecisi
on=-203 (Accessed on 26 November 2011.)

[IEEE 1998]   "IEEE STD 830-1998: IEEE Recommended Practice for Software Requirements
Specifications." Institute of Electrical and Electronics Engineers. Available via:
http://ieeexplore.ieee.org/xpl/freeabs_all.jsp?arnumber=720574 (Accessed on 27
November 2011.)

[ISO 2007]   International Organization for Standardization. "ISO 19131:2007 Geographic
Information – Data product specifications." Available via:
http://www.iso.org/iso/iso_catalogue/catalogue_tc/catalogue_detail.htm?csnumb
er=36760 (Accessed on 9 May 2012.)

[ISO 2009]   International Organization for Standardization. "ISO 31000 – Risk Management
– Principles and Guidelines." Available via:
http://www.iso.org/iso/iso_catalogue/management_and_leadership_standards/ris
k_management.htm. and http://www.iso.org/iso/pressrelease.htm?refid=Ref1266
(Accessed on 2 November 2011.)

[ISO 2011a]   "ISO 9241 Series – Ergonomics of human-system interaction." Available via:
http://www.iso.org/iso/iso_catalogue/catalogue_tc/catalogue_tc_browse.htm?co
mmid=53372 (Accessed on 26 November 2011.)

[ISO 2011b]	"ISO/TR 16982:2002 Ergonomics of human-system interaction – Usability methods supporting human-centred design." Available via: http://www.iso.org/iso/catalogue_detail?csnumber=31176 (Accessed on 26 November 2011.)
[Kalasky 2010]	Kalasky, D., M. Coffmann, M. DeGrano, and K. Field. "Simulation-Based Manpower Planning With Optimized Scheduling in a Distributed Multi-User Environment." In *Proceedings of the 2010 Winter Simulation Conference*, edited by B. Johansson, S. Jain, J. Montoya-Torres, J. Hugan, and E. Yücesan, 3447–3459. Piscataway, New Jersey: Institute of Electrical and Electronics Engineers, Inc. 2010.
[KCOEM 2011]	"Hazardous Material Release." King County Office of Emergency Management. Available via: http://www.kingcounty.gov/safety/prepare/residents_business/Hazards_Disasters/HazardousMaterials.aspx (Accessed on 26 November 2011.)
[Kelton 1998]	Kelton, D., R. Sadowski, and D. Sadowski. *Simulation With Arena*. McGraw-Hill: New York, NY. 1998.
[Knepell 1993]	Knepell, P. and D. Arangno. *Simulation Validation: A Confidence Assessment Methodology (Systems)*. Wiley-IEEE Computer Society Press: Hoboken, NJ. 1993
[Kossiakoff 2003]	Kossiakoff, A. and W. Sweet. *Systems Engineering: Principles and Practice*. John Wiley Sons: Hoboken, NJ. 2003.
[Kotenko 2007]	Kotenko, I. 2007. "Multi-agent Modelling and Simulation of Cyber-Attacks and Cyber-Defense for Homeland Security." *4th IEEE Workshop on Intelligent Data Acquisition and Advanced Computing Systems* (IDAACS 2007).
[Kuhl 2000]	Kuhl, F., R. Weatherly, and J. Dahmann. *Creating Computer Simulation Systems: An Introduction To The High Level Architecture*. Prentice-Hall 2000.
[LANL 2005]	"CartaBlanca: A High-Efficiency, Object-Oriented, General-Purpose Computer Simulation Environment. LAUR-05-6574." Los Alamos National Laboratory, Los Alamos, NM. Available via: http://www.lanl.gov/orgs/tt/pdf/techs/cartablanca_entry.pdf (Accessed on 30 October 2011.)
[Lanner 2011]	Lanner Group Ltd. 2011. "L-SIM – Java component for Business Process Simulation." Available via: http://www.lanner.com/en/l-sim.cfm (Accessed on 7 November 2011.)
[Law 2000]	Law, A. M., W. D. Kelton, 2000. *Simulation modeling and analysis, 3rd edition*. New York: McGraw-Hill.
[Lever 2004]	Lever, A. "Fidelity and Negative Training in System Simulation." Available via: http://www.vega-group.com/assets/documents/10000410fidelity.PDF (Accessed on 23 November 2011.)

[Libicki 1995]   Libicki, M. *Information Technology Standards: Quest for the Common Byte.* Digital Press: Maynard, MA. 1995.

[Liu 2009]   Liu, J., Y. Li, and Y. He. 2009. "A Large-Scale Real-Time Network Simulation Study Using PRIME." In *Proceedings of 2009 Winter Simulation Conferenc.* Edited by M. D. Rossetti, R. R. Hill, B. Johansson, A. Dunkin, and R. G. Ingalls, 797-806. Piscataway, New Jersey: Institute of Electrical and Electronic Engineers.

[Macal 2011]   Macal, C. M. and M. J. North. "Introductory Tutorial: Agent-Based Modeling and Simulation." In *Proceedings of the 2011 Winter Simulation Conference,* edited by S. Jain, R.R. Creasey, J. Himmelspach, K.P. White, and M. Fu, 1456-1469. Piscataway, New Jersey: Institute of Electrical and Electronics Engineers, Inc. 2011.

[Macaulay 1996]   Macaulay, L. *Requirements Engineering.* Springer: London, England. 1996.

[Mackerrow 2006]   Mackerrow, E. "Modeling the Socio-Political Dynamics of Islamist Movements." *26th Annual Conference of the Center for Nonlinear Studies. August 14-18, 2006* – Oppenheimer Center, LANL, Los Alamos, NM, USA. 2006.

[Marshall 2000]   Marshall, C. *Enterprise Modeling with UML: Designing Successful Software Through Business Analysis.* Addison-Wesley: Reading, MA. 2000.

[McLean 2009]   McLean, C.R., S. Jain, and Y.T. Lee. "Overview of MSA Needs for Homeland Security." Interservice/ Industry Training, Simulation, and Education Conference (I/ITSEC) 2009. Paper No. 9505. Orlando, FL. Nov. 30-Dec. 2, 2009.

[McLean 2011a]   McLean, C., Y. T. Lee, S. Jain, C. Hutchings. "Modeling and Simulation of Healthcare Systems for Homeland Security Applications – NISTIR 7784." National Institute of Standards and Technology: Gaithersburg, MD. September 2011. Available via: http://www.nist.gov/customcf/get_pdf.cfm?pub_id=907712 (Accessed on 21 November, 2011.)

[Meitzler 2009]   Meitzler, W.D., S.J. Ouderkirk, and C.O. Hughes. "Security Assessment Simulation Toolkit (SAST)." Pacific Northwest National Laboratory. November 2009. Available via: www.hsdl.org/?view&did=687669 (Accessed on 26 November 2011.)

[Mena 2004]   Mena, J. *Homeland Security: Techniques and Technologies.* Charles River Media: Hingham, MA. 2004.

[MSPCC 2011]   "The Organization for Developing and Providing Professional Certification." Available via: http://www.simprofessional.org/index.html (Accessed on 25 November 2011.)

[NASA 2003]   Moran, Patrick J. "An Open Source Option for NASA Software," NAS Technical Report NAS-03-009, NASA Ames Research Center, Moffett Field

CA, April 2003.

[NASA 2008] "Standard For Models and Simulations – NASA Technical Standard (NASA-STD-7009)." National Aeronautics and Space Administration. Washington, DC. July 2008. Available via: http://standards.nasa.gov/documents/viewdoc/3315599/3315599 (Accessed on 27 November 2011.)

[NDIA 2012] National Defense Industrial Association (NDIA). Inaugural Event of the National Modeling and Simulation Coalition. Washington DC. February 2012. Available via: http://www.ndia.org/meetings/21C0/Pages/default.aspx (Accessed on 16 February 2012.)

[Neuman 2002] Neuman, S.P. and P. J. Wierenga. "A Comprehensive Strategy of Hydrogeologic Modeling and Uncertainty Analysis for Nuclear Facilities and Sites, NUREG/CR-6805." Office of Nuclear Regulatory Research, U. S. Nuclear Regulatory Commission. 2002.

[NIPP 2009] National Infrastructure Protection Plan. U.S. Department of Homeland Security. Available via: http://www.dhs.gov/xlibrary/assets/NIPP_Plan.pdf (Accessed on 22 November 2011.)

[NISAC 2007] National Infrastructure Simulation and Analysis Center (NISAC). "National Population, Economic, and Infrastructure Impacts of Pandemic Influenza with Strategic Recommendations." U.S. Department of Homeland Security. October 2007. Available via: http://www.sandia.gov/nisac/docs/PI_FINAL_1-25-08_unlimited.pdf (Accessed on 30 October 2011.)

[NISAC 2011] National Infrastructure Simulation and Analysis Center (NISAC). "NISAC Tools: SimCore." Available via: http://www.lanl.gov/programs/nisac/simcore.shtml (Accessed on 7 November 2011.)

[NISAC 2011b] "National Infrastructure Simulation and Analysis Center – Fast Analysis and Simulation Team (FAST)." Available via: http://www.sandia.gov/nisac/fast.html (Accessed on 22 November 2011.)

[NIST 2007] "NIST Exercise Control System Wins Approval at Golden Guardian 2007." Available via: http://www.nist.gov/el/msid/11_01_07.cfm (Accessed on 22 November 2011.)

[NIST 2011a] "NIST Publications Portal." Available via: http://www.nist.gov/publication-portal.cfm (Accessed on 27 November 2011.)

[NIST 2011b] DHS/NIST Workshop on Homeland Security Modeling & Simulation, June 14-15, 2011, NISTIR 7826. National Institute of Standards and Technology: Gaithersburg, MD. November 2011.

[NIST 2012] McLean, C., C. Hutchings, S. Jain, Y. T. Lee. Technical Guidance for the Maintenance, Support, and Deployment of Homeland Security Simulation Applications. National Institute of Standards and Technology: Gaithersburg,

MD. NISTIR 7845. March 2012.

[NNSA 2004]   National Nuclear Security Administration (NNSA). "ASC Program Plan, FY06,
                   NA-ASC-106R-05-Vol. 1-Rev. 0." Washington, DC. 2004.
                   https://asc.llnl.gov/publications/asc_program_plan_fy06.pdf (Accessed 05
                   March 2012.)

[NRC 1995]   National Research Council. *Standards, Conformity Assessment, and Trade Into
                   The 21st Century.* National Academy Press, Washington, DC. 1995.

[NRC 2003]   Committee on the Atmospheric Dispersion of Hazardous Material Releases,
                   National Research Council, *Tracking and Predicting the Atmospheric
                   Dispersion of Hazardous Material Releases: Implications for Homeland
                   Security,* National Academy Press, Washington, DC, 2003.

[NWS 2009]   Digital Weather Markup Language Specification. National Weather Service
                   (NWS). Available via:
                   http://www.nws.noaa.gov/mdl/XML/Design/MDL_XML_Design.pdf [accessed
                   May 9, 2012]

[Oberkampf 2010]   Oberkampf, W.L. and C.L. Roy, *Verification and Validation in Scientific
                   Computing,* Cambridge University Press, New York, 2010.

[O'Hara 2010]   O'Hara, S., McLean, C., and Lee, Y. T. "Modeling and Simulation for
                   Emergency Management and Health Care Systems: Workshop Summary –
                   NISTIR 7684." National Institute of Standards and Technology: Gaithersburg,
                   MD. April 2010. Available via:
                   http://www.nist.gov/customcf/get_pdf.cfm?pub_id=905064
                   (Accessed on 21 November 2011.)

[Omerod 2001]   Ormerod, R.J. "Viewpoint: The Success and Failure of Methodologies – a
                   Comment on Connell (2001) Evaluating Soft OR." Journal of the Operational
                   Research Society, 52(10). 1176-1179.

[OMG 2012]   *Documents associated with UML Version 2.2.* Object Management Group.
                   Available on-line via: http://www.omg.org/spec/UML/2.2/ [last accessed on
                   May 9, 2012].

[OR/MS 2011]   Simulation Software Survey. OR/MS Today. Available via: http://www.orms-
                   today.org/surveys/Simulation/Simulation.html (Accessed on 30 October 2011.)

[ORAU 2011]   "Exercise Builder." Oak Ridge Associated Universities (ORAU). Available via:
                   http://orise.orau.gov/emi/training-products/exercisebuilder/default.htm
                   (Accessed on 20 November 2011.)

[ORNL 2006]   Guidelines for Archiving Data in the NARSTO Permanent Data Archive. North
                   American Research Strategy for Tropospheric Ozone (NARSTO). Available
                   via:
                   http://cdiac.ornl.gov/programs/NARSTO/Guidelines_for_Archiving_NARSTO_
                   Data.pdf [accessed May 9, 2012]

[OSI 2011a]   Open Source Initiative. "Mission." Available via: http://www.opensource.org/ (Accessed on 22 October 2011.)

[OSI 2011b]   Open Source Initiative. "Open Source Definition." Available via: http://www.opensource.org/osd.html. (Accessed on 20 November 2011.)

[Pace 2004]   Pace, D.K. "Modeling and Simulation Verification and Validation Challenges." Johns Hopkins APL Technical Digest, Volume 25, Number 2. John Hopkins Applied Physics Laboratory. 2004. Available via: http://www.jhuapl.edu/techdigest/TD/td2502/Pace.pdf (Accessed on 26 November 2011.)

[Patriot 2001] Patriot Act – HR 3162. October 24, 2001. Available via: http://www.gpo.gov/fdsys/pkg/BILLS-107hr3162rds/pdf/BILLS-107hr3162rds.pdf (Accessed on 20 November 2011.)

[PITAC 2000]   "Report to the President: Developing Open Source Software To Advance High End Computing." President's Information Technology Advisory Committee: Panel on Open Source Software for High End Computing. October 2000. *Available via:* http://www.nitrd.gov/Pubs/pitac/pres-oss-11sep00.pdf (Accessed on 21 October 2011.)

[Powersim 2011] Studio 8 Developer Suite: Studio 8 Simulation Engine. Powersim Software. Available via: http://www.powersim.com/main/products___services/powersim_products/simulation-engine/ (Accessed on 10 Nov. 2011.)

[Reilly 2004]   Reilly, T.E. and A. W. Harbaugh. "Guidelines for Evaluating Ground-Water FlowModels - Scientific Investigations Report 2004-5038." U.S. Geological Survey, Washington, DC. 2004.

[Richbourg 2008] Richbourg, R. and R. Lutz. "Live Virtual Constructive Architecture Roadmap (LVCAR) Comparative Analysis of the Architectures." DoD Modeling Simulation Coordination Office (MSCO). Available via: http://www.msco.mil/documents/_19_LVCAR%20-%202%20of%205%20-%20CAA%20-%2020090716.pdf (Accessed on 11 November 2011.)

[Rohrer 1998]   Rohrer, M. and J. Banks. "Required Skills of a Simulation Analyst." IIE Solutions. May 1998, Vol. 30 Issue 5, pp. 20-24. Available via: http://web.ebscohost.com/ehost/detail?sid=06b960e9-892c-4ba5-82ce-3914ee3e993f%40sessionmgr15&vid=1&hid=10&bdata=JnNpdGU9ZWhvc3QtbGl2ZQ%3d%3d#db=f5h&AN=623277 (Accessed on 26 November 2011.)

[Rumbaugh 1999] Rumbaugh, J., I. Jacobsen, and Booch, G. *The Unified Modeling Language Reference Manual*. Addison-Wesley, Reading, MA. 1999.

[Salen 2004]   Salen, K. and E. Zimmerman. *Rules of Play: Game Design Fundamentals.* MIT Press: Cambridge, MA 2004.

[Sandia 1998]   "Strategic Computing & Simulation Validation & Verification Program –

Program Plan." Sandia National Laboratories. April 1998. Available via: http://www.sandia.gov/asc/pubs_pres/pubs/vnvprogplan_FY98.html (Accessed on 26 November 2011.)

[Schaffer 2004]   Schaffer, E.M. "How to Develop an Effective GUI Standard." Human Factors International. Fairfield, IA. 2004. Available via: www.humanfactors.com/downloads/guistandards.pdf (Accessed on 12 November 2011.)

[Schriber 1998]   Schriber, T. and D. Bruner. "How Discrete Event Simulation Works." In *Handbook of Simulation: Principles, Methodology, Advances, Applications, and Practice.* Edited by J. Banks. Wiley-Interscience: New York, NY. 1998.

[Schriber 2011]   Schriber, T. J., and D. T. Brunner. 2011. "Inside discrete-event simulation software: How it works and why it matters." In *Proceedings of 2011 Winter Simulation Conference,* edited by S. Jain, R.R. Creasey, J. Himmelspach, K.P. White, and M. Fu. Piscataway, New Jersey: Institute of Electrical and Electronic Engineers.

[Schwetman 2001]   Schwetman, H. 2001. "CSIM19: A Powerful Tool For Building System Models." In Proceedings of 2001 Winter Simulation Conference, edited by B. A. Peters, J. S. Smith, D. J. Medeiros, and M. W. Rohrer. Piscataway, New Jersey: Institute of Electrical and Electronics Engineers.

[SDS 2011]   Annual international conference of the System Dynamics Society. Available via: http://conference.systemdynamics.org/ (Accessed on 20 October 2011.)

[Section508 2011]   "Resources for understanding and implementing Section 508." Available via: http://www.section508.gov/index.cfm (Accessed on 26 November 2011.)

[Senate 2007]   "Public Law 110–53—Aug. 3, 2007 Implementing Recommendations of the 9/11 Commission Act Of 2007." U.S. Senate, Washington, DC. 2007. Available via: http://intelligence.senate.gov/laws/pl11053.pdf (Accessed on 26 November 2011.)

[Shannon 1975]   Shannon, R.E., *Systems Simulation: The Art and Science*, Prentice-Hall, 1975.

[Shaw 1996]   Shaw, M., and D. Garlan. *Software Architecture: Perspectives on an Emerging Discipline.* Prentice-Hall: Saddle River, NJ. 1996.

[Shendarkar 2008]   Shendarkar, A., K. Vasudevan, S. Lee, Y. Son. "Crowd Simulation for Emergency Response using BDI Agents Based on Immersive Virtual Reality." *Simulation Modelling Practice and Theory* 16:1415-1429. 2008.

[SimSummit 2011]   "SimSummit." Available via: http://www.sim-summit.org/Sim_Summit/default.htm (Accessed on 25 November 2011.)

[SISO 2010]   *Standard for Commercial-off-the-shelf Simulation Package Interoperability Reference Models (SISO-STD-006-2010).* Simulation Interoperability Standards Organization, Orlando, Florida.

[SISO 2011]   "GM-VV PDG - Generic Methodology for Verification and Validation."
                   Simulation Interoperability Standards Organization (SISO). Available via:
                   http://www.sisostds.org/StandardsActivities/DevelopmentGroups/GMVVPDGG
                   enericMethodologyforVVAintheM.aspx (Accessed on 20 November 2011.)

[Smith 1986]   Smith, S.L. and J.N. Mosier "Guidelines For Designing User Interface Software
                   (ESD-TR-86-278)." Mitre Corporation. Bedford, MA. Available via:
                   http://www.userlab.com/Downloads/Smith_Mosier_guideline_.pdf (Accessed
                   on 12 November 2011.)

[Sodhi 1992]   Sodhi, J. *Software Requirements Analysis and Specification.* McGraw-Hill: New
                   York, NY. 1992.

[Sommerville 2004] Sommerville, I. *Software Engineering (7th edition).* Pearson/Addision-Wesley:
                   Boston, MA. 2004.

[SourceForge 2011] SourceForge. "Find, Create, and Publish Open Source software for free."
                   Available via: http://sourceforge.net/ (Accessed on 20 November 2011.)

[SQLTutorial 2011] "SQL Tutorial." Available via: http://www.sqltutorial.org/ (Accessed on 20
                   November 2011.)

[Steinhauser 2008] Steinhauser, M.O. *Computational Multiscale Modeling of Fluids and Solids:
                   Theory and Applications.* Springer-Verlag Berlin: Heidelberg, Germany. 2008.

[Sterman 2000]   Sterman, J. *Business Dynamics: Systems Thinking and Modeling for Complex
                   World.* Irwin McGraw-Hill: Boston, MA. 2000.

[StrategyWorld.com DoD Training with Simulations Handbook, Chapter 1. Strategy Page:
2011]   Professional Wargames Page. Available via:
                   http://www.strategypage.com/prowg/simulationshandbook/default.asp
                   (Accessed on 4 November 2011.)

[SUMMIT 2011]   "Overview. Standard Unified Modeling, Mapping, and Integration Toolkit
                   (SUMMIT)." Available via: https://dhs-summit.us/overview-mdo.html.
                   (Accessed on 22 October 2011.)

[Taylor 2003]   Taylor, S., Robinson, S., and Ladbrook, J. "Towards Collaborative Simulation
                   Modelling: Improving Human-To-Human Interaction Through Groupware."
                   Proceedings of the 17th European Simulation Multiconference (ESM 2003), (ed.
                   D. Al-Dabass) October 27-29, 2003. University of Naples II, Naples, Italy.

[Taylor 2011]   Taylor, S. J. E., M. Ghorbani, T. Kiss, D. Farkas, N. Mustafee, S. Kite, S. J.
                   Turner, and S. Strassburger. 2011. "Distributed Computing and Modeling &
                   Simulation: Speeding up Simulations and Creating Large Models." In
                   Proceedings of the 2011 Winter Simulation Conference, edited by S. Jain, R.R.
                   Creasey, J. Himmelspach, K.P. White, and M. Fu, 161-175. Piscataway, New
                   Jersey: Institute of Electrical and Electronic Engineers.

[Thalman 2008]   Thalmann, D. and S.R. Musse. *Crowd Simulation.* Springer-Verlag: London,

England. 2007.

[Trinkle 2011]   Trinkle, J. 2011. "Survey of simulation packages allowing unilateral contacts."
                     Available via: http://www.cs.rpi.edu/~trink/sim_packages.html (Accessed on 7
                     November 2011.)

[USCode 2011]   "Section 3542. Definitions – Information Security." Available via:
                     http://www.law.cornell.edu/uscode/44/3542.html (Accessed on 26 November
                     2011.)

[USLegal 2011]   "Professional Certification Law & Legal Definition." Available via:
                     http://definitions.uslegal.com/p/professional-certification/ (Accessed on 26
                     November 2011.)

[Webopedia 2011a]   "OMG IDL." Available via:
                     http://www.webopedia.com/TERM/O/OMG_IDL.html (Accessed on 21
                     November 2011.)

[Webopedia 2011b]   "Open Architecture." Available via:
                     http://www.webopedia.com/TERM/O/open_architecture.html (Accessed on 20
                     November 2011.)

[WHO 2007]   "Team Building." World Health Organization, Geneva, Switzerland. 2007.
                     Available via: www.who.int/entity/cancer/modules/Team%20building.pdf
                     (Accessed on 26 November 2011.)

[Wieringa 1996]   Wieringa, R. *Requirements Engineering: Frameworks for Understanding*. John
                     Wiley and Sons: Chichester, England. 1996.

[Wiki 2011a]   "Mannequin." Available via: http://en.wikipedia.org/wiki/Mannequin
                     (Accessed on 21 November 2011.)

[Wiki 2011b]   "Mod (video gaming)." Available via: http://en.wikipedia.org/wiki/Game_mod
                     (Accessed on 25 November 2011.)

[Wiki 2011c]   "List of system dynamics software." Available via:
                     http://en.wikipedia.org/wiki/List_of_system_dynamics_software (Accessed on
                     7 November 2011.)

[Wiki 2011d]   "Usability." Available via: http://en.wikipedia.org/wiki/Usability#ISO_9241
                      (Accessed on 21 November 2011.)

[Wiki 2011e]   "Information Model." Available via:
                     http://en.wikipedia.org/wiki/Information_model (Accessed on 20 November
                     2011.)

[Wiki 2011f]   "Data dictionary." Available via: http://en.wikipedia.org/wiki/Data_dictionary
                     (Accessed on 21 November 2011.)

[Wiki 2011g]   "IEEE 829 Standard for Software Test Documentation." Available via:
                     http://en.wikipedia.org/wiki/IEEE_829 (Accessed on 26 November 2011.)

[Worth 2010]   Worth, T., R. Uzsoy, E. Samoff, A.-M. Meyer, J.-M. Maillard, and A. Wendelboe. "Modeling the Response of a Public Health Department to Infectious Disease." In *Proceedings of the 2010 Winter Simulation Conference*, edited by B. Johansson, S. Jain, J. Montoya-Torres, J. Hugan, and E. Yücesan, 2185–2198. Institute of Electrical and Electronics Engineers, Inc.: Piscataway, NJ. 2010.

[WSC 2011]   Annual international Winter Simulation Conference. Available via: www.wintersim.org (Accessed on 30 October 2011.)

[Younker 2002]   Younker, L. W. 2002. "High Tech Help for Fighting Wildfires." Science and Technology Review. Lawrence Livermore National Laboratory. Available via: https://www.llnl.gov/str/November02/pdfs/11_02.pdf (Accessed on 30 October 2011.)

[Zeigler 2000]   Zeigler, B., T. G. Kim, H. Praehofer. *Theory of Modeling and Simulation* (2nd ed.). Academic Press: New York, NY. 2000.

Appendix – DHS M&S Capabilities

This appendix presents a list of M&S capabilities currently used by DHS and an overview of these capabilities. Facilities in this appendix may often be sources (i.e., developers) or recipients of M&S homeland security applications. Other likely recipients of deployed homeland security M&S applications include other Federal government agencies, state and local governments, first responder organizations, and various non-governmental organizations (e.g., healthcare institutions, the Red Cross, and faith-based organizations).

Name	Overview
DHS Operations Battle Lab	Mission: To act as the operational DHS and Interagency experimentation and proof of concept center with the express purpose of enhancing and improving current capabilities, situational awareness, and information sharing by leveraging cost effective existing and emerging technologies, concepts, and processes. Goals: • Transform information sharing and reporting by incorporating a single query federated search capability. • Enhance DHS, federal, state and local (S/L), tribal, and private sector access to data, information, and analysis. • Provide DHS, federal, S/L, tribal, and private sector authorities access to an advanced analytical toolkit to support data correlation and fusion. • Enhance interoperability with DoD, U.S. Northern Command (USNORTHCOM), and National Guard Bureau (NGB) during crisis operations. • Enhance situational awareness by integrating interagency, incident command (IC), law enforcement, proprietary, and multiple classification data into a user-defined picture and share via a role and privilege-based interface. • Provide real time data to facilitate timely, risk-mitigated decision making.
Homeland Security Studies and Analysis Institute (HSSAI)	HSSAI is a Federally Funded Research and Development Center (FFRDC) established by the Homeland Security Act of 2002. HSSAI delivers independent and objective analyses and advises in core areas important to all DHS components in support of policy development, decision making, analysis of alternative approaches, and evaluation of new ideas on issues requiring scientific, technical, and analytical expertise. HSSAI efforts are framed around nine strategic capability areas: • Risk analysis – at the strategic, tactical, and operational levels. • Operations analysis – to improve real-world processes • Threat analysis – identify and understand existing and emerging threats. • Systems analysis – illuminate complex interdependencies and tradeoffs. • Information-sharing analysis – improve the effectiveness and efficiency of HS operations among all levels of government and the

	private sector.
	• Policy and planning analysis – help set the direction for homeland security.
	• Program analysis – identify solutions for capability gaps.
	• Science and technology analysis – ensure advances in these areas benefit the end users.
	• Training, education, and professional development analysis – ensure the necessary competencies for the homeland security workforce.
Homeland Security Systems Engineering and Development Institute (SEDI)	The SEDI FFRDC was established in early 2009 to provide systems engineering expertise and acquisition strategy advice to improve policies, processes, and tools for mission capabilities that ensure the nation's security. SEDI provides an interdisciplinary engineering approach to the challenges of homeland security, combining technical expertise, domain knowledge, and business capabilities to improve interoperability, develop flexible and expandable architectures, and integrate proven technology into practical solutions. Key areas for SEDI work include: • Border security and immigration. • Intelligence and cyber analysis. • Preparedness, response, and recovery. • Protection of critical infrastructure. • Screening and credentialing. • Transportation security.
National Exercise Simulation Center (NESC)	The NESC provides a state-of-the-art facility at FEMA to serve the all-hazards preparedness and response mission through pooling resources, maximizing efficiency, and providing sustained exercise and training support to all stakeholders. NESC will expand its capabilities to: • Support national, federal, state, and local exercises throughout the United States and internationally, with around-the-clock availability, to include Radiological Emergency Preparedness Program exercises and the National Level Exercise with Master Control Cell and National Simulation Cell and related *functions*. • Provide a forum for interagency planners to test their plans (e.g., annual hurricane plans, pandemic influenza plans) by providing realistic incident scenarios through which partners can identify gaps and determine courses of action. • Serve as "future planning" support for FEMA's Disaster Operations Directorate and other FEMA and DHS Directorates by providing technical modeling and simulation tools that enable planners to better visualize potential future scenarios. • Coordinate activities that support real-world events and exercises, such as Homeland Security Exercise and Evaluation Program training and initial-, mid-, and final- planning conferences. • Incorporate real-time, mock-media capabilities, such as the Virtual News Network, that provide exercise participants with realistic breaking news bulletins, interviews, and live news coverage of incidents. • Provide practical training opportunities to those learning about exercise design, conduct, and management through individual mentorship of federal, state, tribal, and local professionals.

	• Link to other centers that provide specialty modeling, simulation and data services such as the Joint War-fighting Center, the National Infrastructure Simulation and Analysis Center, the Emergency Management Institute, and other centers that provide natural and man-made disaster modeling, constructive simulation tools, and Subject Matter Expert (SME) databases and historical deployment/response information. Incorporate national and FEMA improvement management services to include the Remedial Action Management Program, the National Corrective Action Program, and the Lessons Learned Information System, to support in-depth analysis of real-world and exercise events and provide real-time lessons learned capabilities
Office of Infrastructure Protection- Homeland Infrastructure Threat and Risk Analysis Center (HITRAC)/National Infrastructure Simulation and Analysis Center (NISAC)	HITRAC performs threat, vulnerability, and consequence analysis for risk to the National Critical Infrastructure and the capabilities that infrastructure provides. The Modeling, Simulation, and Analysis capability is directed and managed by HITRAC and executed by NISAC. NISAC provides strategic, multi-disciplinary analyses of interdependencies and the consequences of infrastructure disruptions across all 18 critical infrastructure and key resource (CIKR) sectors at National, regional, and local levels. HITRAC manages the capability development and employment of tools for HITRAC and NISAC analysts to address the complexities of interdependent national infrastructures, including process-based systems dynamics models, mathematical network optimization models, physical-science-based models of existing infrastructures, and high-fidelity agent-based simulations of systems.
Office of Intelligence and Analysis (I&A) "Future Series" Workshops	The Strategic Analysis Group has implemented a pilot program of Future Workshops as a follow on the Homeland Security Threat Assessment (HSTA). These workshops have two principal goals. The first is to better integrate intelligence analysts with planning and programming analysts across the DHS Enterprise to support capabilities and gaps analysis. The second is to identify Emerging Trends that may impact homeland security, a requirement of the 9/11 Commission emphasis on preventing surprise. The topics of these workshops are derived from the HSTA Threat Streams, but are focused on more detailed levels of analysis of interest to the Policy customer. These workshops have been jointly hosted by the Undersecretary for Intelligence and Analysis and the Undersecretary for Policy.
Office of Risk Management & Analysis (RMA)	RMA was established in April 2007 and delegated its authority (Delegation Number: 17001) to lead DHS's efforts to establish a common framework to address the overall management and analysis of homeland security risk, develop systematic rigorous risk analysis methodologies, and provide core risk-analysis capabilities to be used throughout DHS to enhance homeland security risk management. RMA addresses risks to the Nation from the DHS Enterprise and the broader Homeland Security Enterprise perspectives. RMA leads DHS's efforts to develop a framework and embed a consistent coordinated approach to address the overall management and analysis of homeland

	security risk. RMA partners with DHS component organizations to develop and apply systematic risk analysis methodologies and to ensure risk information is used effectively to manage homeland security risk. Additionally, RMA is developing and implementing cross-component analysis in the Risk Assessment Process for Informed Decision Making (RAPID) and in the Homeland Security Nation Risk Assessment (HSNRA). The risk information produced through these assessments will inform enterprise level strategic planning and resource allocation processes and decisions.
Science and Technology Directorate (S&T) -- Operations Analysis Division	The Director of the Operations Analysis Division (OAD) is responsible to the S&T Under Secretary as the principal assistant for operations analysis and senior advisor on requirements for operations analysis, initiating and conducting projects as required and overseeing and managing analytical resources within the S&T Directorate. The OAD director is executive agent of the two DHS FFRDCS (HSSAI and SEDI), as well as executive director of the Homeland Security Science and Technology Advisory Committee (HSSTAC). OAD also manages two other programs with cross-Component applicability: • Gaming Simulations conduct seminar games focused on resolving technology transition challenges. These games result in operational capability gap articulation and understanding, as well as identifying potential technology solutions. Games conducted by OAD have helped identify organizational impediments to homeland security technology handoff and fielding. Games typically involve participation from multiple DHS Components, private sector, technology experts, and interagency personnel. • Operational Experimentation assists DHS components by conducting experiments in real-world environments that involve actual users and operators (vice scientists and engineers) and a realistic mix of operational systems in addition to those being tested. Operational experiments are focused on the effects of systems on operations, as opposed to test and evaluation, which tends to consider the effects of operations upon systems.
Science and Technology Directorate (S&T) - University Programs Centers of Excellence	The Homeland Security Act of 2002 granted DHS the authority to create university-based Centers of Excellence (COEs). The centers are authorized by Congress and chosen by the S&T through a competitive selection process. "The Secretary, acting through the Under Secretary for Science and Technology, shall designate a university-based center or several university-based centers for homeland security. The purpose of the center or these centers shall be to establish a coordinated, university-based system to enhance the Nation's homeland security." – as amended. The COEs bring together leading experts and researchers to conduct multidisciplinary research and education for homeland security solutions. Each center is led by a university in collaboration with partners from other institutions, agencies, laboratories, think tanks, and the private sector. Current COEs include: • The Center for Risk and Economic Analysis of Terrorism Events (CREATE), led by the University of Southern California, develops advanced tools to evaluate the risks, costs and consequences of

| | terrorism, and guides economically viable investments in countermeasures that will make our Nation safer and more secure.
- The National Center for Foreign Animal and Zoonotic Disease Defense (FAZD), led by Texas A&M University, protects against the introduction of high-consequence foreign animal and zoonotic diseases into the United States, with an emphasis on prevention, surveillance, intervention, and recovery.
- The National Center for Food Protection and Defense (NCFPD), led by the University of Minnesota, defends the safety and security of the food system from pre-farm inputs through consumption by establishing best practices, developing new tools, and attracting new researchers to prevent, manage, and respond to food contamination events
- The National Consortium for the Study of Terrorism and Responses to Terrorism (START), led by the University of Maryland, makes decisions on how to disrupt terrorists and terrorist groups, while strengthening the resilience of U.S. citizens to terrorist attacks.
- The Center for Advancing Microbial Risk Assessment (CAMRA), led by Michigan State University, Drexel University, and established jointly with the U.S. Environmental Protection Agency, fills critical gaps in risk assessments for decontaminating microbiological threats, such as plague and anthrax, and answering the question, "How Clean is Safe?"
- The National Center for the Study of Preparedness and Catastrophic Event Response (PACER) led by Johns Hopkins University, optimizes our nation's preparedness in the event of a high-consequence natural or man-made disaster, as well as develops guidelines to best alleviate the effects of such an event.
- The Center of Excellence for Awareness & Location of Explosives-Related Threats (ALERT), led by Northeastern University in Boston, MA and the University of Rhode Island in Kingston will develop new means and methods to protect the nation from explosives-related threats, focusing on detecting leave-behind Improvised Explosive Devices, enhancing aviation cargo security, providing next-generation baggage screening, detecting liquid explosives, and enhancing suspicious passenger identification.
- The National Center for Border Security and Immigration (NCBSI), led by the University of Arizona in Tucson (research co-lead) and the University of Texas at El Paso (education co-lead), is developing technologies, tools, and advanced methods to balance immigration and commerce with effective border security, as well as assess threats and vulnerabilities, improve surveillance and screening, analyze immigration trends, and enhance policy and law enforcement efforts.
- The Center for Maritime, Island and Port Security (MIPS), led by the University of Hawaii in Honolulu for maritime and island security and Stevens Institute of Technology in Hoboken, New Jersey, for port security, will strengthen maritime domain awareness and safeguard populations and properties unique to U.S. islands, ports, and remote and extreme environments.
- The Center for Natural Disasters, Coastal Infrastructure, and Emergency Management (NDCIEM), led by the University of North Carolina at Chapel Hill and Jackson State University in Jackson, Mississippi will enhance the Nation's ability to safeguard populations, properties, and economies as it relates to the |
|---|---|

	consequences of catastrophic natural disasters.
• The National Transportation Security Center of Excellence (NTSCOE), was established in accordance with HR1, Implementing the Recommendations of the 9/11 Commission Act of 2007, in August 2007. NTSCOE is made up of seven institutions: Connecticut Transportation Institute at the University of Connecticut, Tougaloo College, Texas Southern University, National Transit Institute at Rutgers - The State University of New Jersey, Homeland Security Management Institute at Long Island University, Mack Blackwell National Rural Transportation Study Center at the University of Arkansas, and the Mineta Transportation Institute at San José State University. The NTSCOE will develop new technologies, tools and advanced methods. The goal is to defend, protect, and increase the resilience of the nation's multi-modal transportation infrastructure and education and training baselines for transportation security geared towards transit employees and professionals.
• The Center of Excellence in Command, Control and Interoperability (C2I), led by Purdue University (visualization sciences co-lead) and Rutgers University (data sciences co-lead), will create the scientific basis and enduring technologies needed to analyze massive amounts of information from multiple sources to more reliably detect threats to the security of the nation and its infrastructures, and to the health and welfare of its populace. These new technologies will also improve the dissemination of both information and related technologies.

www.ingramcontent.com/pod-product-compliance
Lightning Source LLC
Chambersburg PA
CBHW080304180526
45167CB00006B/2658